A BRIEF INTRODUCTION TO SYMPLECTIC AND CONTACT MANIFOLDS

NANKAI TRACTS IN MATHEMATICS

Series Editors: Yiming Long, Weiping Zhang and Lei Fu
Chern Institute of Mathematics

Published*

*For the complete list of titles in this series, please visit
http://www.worldscientific.com/series/ntm

Nankai Tracts in Mathematics – Vol. 15

A BRIEF INTRODUCTION TO SYMPLECTIC AND CONTACT MANIFOLDS

Augustin Banyaga

The Pennsylvania State University, USA

Djideme F Houenou

Institute of Mathematics and Physical Sciences, Benin

 **World Scientific**

NEW JERSEY · LONDON · SINGAPORE · BEIJING · SHANGHAI · HONG KONG · TAIPEI · CHENNAI · TOKYO

Published by

World Scientific Publishing Co. Pte. Ltd.

5 Toh Tuck Link, Singapore 596224

USA office: 27 Warren Street, Suite 401-402, Hackensack, NJ 07601

UK office: 57 Shelton Street, Covent Garden, London WC2H 9HE

Library of Congress Cataloging-in-Publication Data

Names: Banyaga, Augustin. | Houenou, Djideme F.

Title: A brief introduction to symplectic and contact manifolds / by Augustin Banyaga
 (The Pennsylvania State University, USA), Djideme F. Houenou
 (International Centre for Theoretical Physics, Italy).

Description: New Jersey : World Scientific, 2016. | Series: Nankai tracts in mathematics ; 15 |
 Includes bibliographical references and index.

Identifiers: LCCN 2016025065 | ISBN 9789814696708 (hardcover : alk. paper)

Subjects: LCSH: Symplectic and contact topology. | Mathematical physics. |
 Symplectic manifolds. | Contact manifolds.

Classification: LCC QA613.659 .B36 2016 | DDC 516.3/6--dc23

LC record available at https://lccn.loc.gov/2016025065

British Library Cataloguing-in-Publication Data

A catalogue record for this book is available from the British Library.

Printed in Singapore

Introduction

This book is a brief introduction to the study of symplectic manifolds and their odd dimensional analogs: the contact manifolds.

Chapter 1 deals with the linear theory. From the very start, we point out the connection between linear symplectic structures with inner product and complex structures. We prove that all symplectic structures on vector spaces of the same dimension are isomorphic (existence of canonical basis).

In chapter 2, we give the definition of symplectic manifolds and provide some basic examples. We prove Darboux theorem using Moser path method. Then we initiate the study of symplectomorphisms. We give a general method for constructing them which in turn is used to show that the group of symplectomorphisms acts p-transitively on connected symplectic manifolds (Boothby Theorem). We mention that this "infinite dimensional Lie group" determines the symplectic geometry. We also introduce the important Calabi homomorphism. Particular attention is paid to Lagrangian submanifolds, a unifying notion of many things in symplectic geometry.

In chapter 3, we pay attention to the particular case of Hamiltonian diffeomorphisms and introduce the notion of integrable systems. We also introduce the Poisson brackets and more generally the notion of Poisson manifolds.

In chapter 4, we introduce group actions on manifolds and specifically hamiltonian and symplectic actions on symplectic manifolds. We introduce the notion of momentum map and state the famous convexity theorem of Atiyah, Guillemin-Sternberg for the momentum map of a torus hamiltonian action and the Marsden-Weinstein reduction theorem.

Chapter 5 is an introduction to contact manifolds. It contains the basic definitions: Reeb field, contact dynamics, Darboux theorem for contact manifolds and a good introduction to relations between symplectic and contact structures. Moreover, we present some basic theorems like the prequantization bundle of integral symplectic manifolds (Boothby-Wang theorem) and the stability theorem of the contact structures (Gray-Martinet theorem).

In Appendix A we give a review of differential forms and the de Rham cohomology. The reader is invited to go there when he/she needs it.

In chapter 6, we collected solutions to a few selected exercises.

The material in chapter 1 to 6 is elementary and can be used for a one semester first year graduate course. Actually, we used Chapter 1 and 2 for a five weeks course on the introduction to symplectic geometry for a first year graduate course at the Institute of Mathematics and Physical Sciences (IMSP) in Porto Novo (Benin).

The Epilogue (Chapter 7) summarizes the ideas around the interplay between symplectic and contact geometries on one hand, and the uniform topology on the other hand. This is a very active area of nowadays research.

One looks for those symplectic or contact notions which survive uniform limits. For instance, we prove the famous Eliashberg-Gromov rigidity theorem: *"if the uniform limit of a sequence of symplectic diffeomorphisms is a smooth diffeomorphism then the limit is a symplectic diffeomorphism"*.

In Appendix B we give a comprehensive introduction to the study of completely integrable systems in contact geometry following an unpublished paper by the first author and Pierre Molino at the University of Montpellier, (with our many thanks to him). In this chapter we review the theory of completely integrable hamiltonian system, and their generalization. We give the contact analog of contact moment map, state and proof the contact analog of the convexity theorem of Atiyah, Guillemin-Sternberg for the contact moment map.

The Epilogue and Appendix B are more advanced and can serve as a reference for mathematicians or as an introduction to current research.

We would like to express our gratitude to our sponsors: the International Centre for Theoretical Physics (ICTP), Trieste, the African Center for Excellence in Mathematics and its Applications at IMSP and the IMSP for its hospitality while we were preparing this book.

Porto Novo, February 1, 2016

Augustin Banyaga   Djideme Franck Houenou

Contents

Symplectic vector spaces

1.1 Bilinear forms

A **bilinear form** on a (real) vector space V is a map $b : V \times V \longrightarrow \mathbb{R}$ which is linear in each variable, i.e:

$$
\begin{aligned}
b(u + u', v) &= b(u, v) + b(u', v) & \forall\, u, u', v \in V \qquad (1.1.1) \\
b(\lambda u, v) &= \lambda\, b(u, v) & \forall\, u, v \in V, \quad \lambda \in \mathbb{R}
\end{aligned}
$$

and analogously in v.

A bilinear form is symmetric if $b(u, v) = b(v, u)$ and antisymmetric if

$$b(u, v) = -b(v, u).$$

A bilinear form $b : V \times V \longrightarrow \mathbb{R}$ determines a linear map $\tilde{b} : V \longrightarrow V^*$ (where V^* is the dual of V, i.e. the space of linear maps $V \longrightarrow \mathbb{R}$), by

$$
\begin{aligned}
\tilde{b}(u) : \quad V &\longrightarrow \mathbb{R} \\
v &\longmapsto \tilde{b}(u)(v) = b(u, v).
\end{aligned}
$$

The rank of b is the dimension of the image of $\tilde{b}$. The bilinear form b is said to be non-degenerate if $\tilde{b}$ is an isomorphism.

Definition 1.1

A **symplectic form** *on a vector space V is a non-degenerate antisymmetric bilinear form $\omega : V \times V \longrightarrow \mathbb{R}$.*

The couple (V, ω) of a vector space and a symplectic form on V is called a **symplectic vector space.**

1.2 Basis

If we fix a basis $\mathcal{E} = (e_1, \cdots, e_{2n})$ of the vector space V, then any bilinear form $b : V \times V \longrightarrow \mathbb{R}$ can be represented by a matrix $M_b = (\alpha_{ij})$ where $\alpha_{ij} = b(e_i, e_j)$.

1

If b is symmetric, then M_b is a symmetric matrix, i.e $^tM_b = M_b$ (where $^tM_b = (\alpha_{ji})$ stands for the transpose of M_b).

If b is antisymmetric, then M_b is a skew-symmetric matrix, i.e $^tM_b = -M_b$ or $\alpha_{ji} = -\alpha_{ij}$, $\forall\ i \neq j$.

The bilinear form b is non-degenerate if and only if M_b is an invertible matrix, i.e. the determinant of M_b is different of zero ($\det M_b \neq 0$).

Hence the matrix M_ω of a symplectic form satisfies:

1. $^tM_\omega = -M_\omega$

2. $\det M_\omega \neq 0$.

Proposition 1.1
   Every symplectic vector space (V, ω) is even dimensional.

   Proof
   We have $\det(^tM_\omega) = \det(M_\omega)$ and $\det(-M_\omega) = (-1)^m \det(M_\omega)$ where $m = \dim(V)$. Equality (1) implies that $\det M_\omega = (-1)^m \det(M_\omega)$ therefore $\det(M_\omega) = 0$ if m is odd.   □

Let us note the following definitions:

1. A **scalar product** on a vector space V is a symmetric bilinear form

$$g : V \times V \longrightarrow \mathbb{R}$$

   which is positive definite.

   This means that if $g(u, v) = 0$ for all $v \in V$ then $u = 0$.

2. A **complex structure** on a vector space V is a linear map

$$J : V \longrightarrow V$$

   such that

$$J\big(J(v)\big) = -v, \qquad \text{for all} \qquad v \in V, \quad \text{i.e} \qquad J^2 = -I.$$

3. A **hermitian structure** on a vector space V is a pair (g, J) where g is a scalar product on V and J a complex structure on V such that

$$g(u, v) = g(Ju, Jv), \qquad \text{for all} \qquad u, v \in V.$$

If we are given a complex structure J and a scalar product g_0, we define another scalar product

$$g(u,v) = \frac{1}{2}\Big(g_0(u,v) + g_0(Ju, Jv)\Big).$$

Then (g, J) is a hermitian structure on V.

Theorem 1.1

Let (g, J) be a hermitian structure on a vector space V. Then

$$\omega : V \times V \longrightarrow \mathbb{R}$$

defined by

$$\omega(u,v) = g(u, Jv)$$

is a symplectic form.

Proof

It is obvious that ω is a bilinear form. Let us prove that it is antisymmetric.

$$
\begin{aligned}
\omega(u,v) = g(u, Jv) &= g\big(Ju, J(Jv)\big) \qquad \text{(hermitian property)}\\
&= g(Ju, -v)\\
&= -g(Ju, v)\\
&= -g(v, Ju)\\
&= -\omega(v, u). \qquad\qquad (1.2.1)
\end{aligned}
$$

The form ω is non-degenerate. Indeed if $u \in \ker(\tilde\omega)$ then $\tilde\omega(u)(v) = 0$ for all $v \in V$, i.e

$$0 = \tilde\omega(u)(v) = \omega(u,v) = g(u, Jv) = -g(Ju, v) \qquad \forall\, v \in V,$$

which implies that $Ju = 0$; therefore $u = 0$ since J is an isomorphism. $\square$

If V is any vector space and $l,\, l' : V \longrightarrow \mathbb{R}$ are two linear maps, we can define

$$
\begin{aligned}
b: \quad \mathbb{R} \times \mathbb{R} &\longrightarrow \mathbb{R}\\
(u,v) &\longmapsto b(u,v) = l(u)l'(v) - l'(u)l(v).
\end{aligned}
$$

This defines an antisymmetric bilinear form $l \wedge l'$ on V.

If $\mathscr{E} = (e_1, \cdots, e_{2n})$ is a basis of V and $(\varepsilon_1, \cdots, \varepsilon_{2n})$ the dual basis, i.e each ε_j is a linear form on V such that

$$\varepsilon_j(e_i) = \begin{cases} 1 & \text{if } i = j \\ 0 & \text{otherwise} \end{cases}$$

then $\varepsilon_i \wedge \varepsilon_j$ are bilinear forms.

If we denote by $\Lambda^2 V$ the space of all bilinear antisymmetric forms on V, we see that $\{\varepsilon_i \wedge \varepsilon_j\}_{i<j}$ form a basis of $\Lambda^2 V$.

Example 1.1

Choosing a basis on a vector space V of dimension m identifies it with $\mathbb{R}^m$. We are now going to consider examples in $\mathbb{R}^m$.

1. *On $\mathbb{R}^m$ with its standard basis where each element $u = (x_1, \cdots, x_m)$ is an $m-$tuple of real numbers, we have the standard scalar product:*

$$g(u, v) = u \cdot v = x_1 x_1' + \cdots + x_m x_m'$$

   denoting $u = (x_1, \cdots, x_m)$ and $v = (x_1', \cdots, x_m')$.

2. *On $\mathbb{R}^{2n}$ where $u = (x_1, \cdots, x_n, x_{n+1}, \cdots x_{2n})$. We note $X = (x_1, \cdots, x_n)$ and $Y = (x_{n+1}, \cdots x_{2n})$ and write $u = (X, Y)$.*

   The map

$$J: \quad \begin{array}{ccc} \mathbb{R}^{2n} & \longrightarrow & \mathbb{R}^{2n} \\ (X, Y) & \longmapsto & J(X, Y) = (Y, -X) \end{array}$$

   is a complex structure, i.e. $J^2 = -I$.

   Also (g, J) is an hermitian structure on $\mathbb{R}^{2n}$. Hence, by Theorem 1.1

$$\omega(u, v) = g(u, Jv) = (X, Y) \cdot (Y', -X') = XY' - X'Y$$

   where $u = (X, Y)$ and $v = (X', Y')$ is a symplectic form on $\mathbb{R}^{2n}$ which *is called the* **standard symplectic form.** *This can be written*

$$\omega(u_1, \cdots, u_{2n}, v_1, \cdots, v_{2n})$$
$$= v_1 u_{n+1} + \cdots + v_n u_{2n} - u_1 v_{n+1} - \cdots - u_n v_{2n}$$
$$= \sum_{j=1}^{n} (v_j u_{n+j} - u_j v_{n+j})$$
$$= \sum_{j=1}^{n} (\varepsilon_j \wedge \varepsilon_{n+j})(u_1 \cdots u_{2n}, v_1 \cdots v_{2n}); \qquad (1.2.2)$$

i.e

$$\omega = \sum_{j=1}^{n} \varepsilon_j \wedge \varepsilon_{n+j}.$$

The matrix of ω with respect to the natural basis of $\mathbb{R}^{2n}$ is

$$
\begin{pmatrix} 0_n & I_n \\ -I_n & 0_n \end{pmatrix}
\equiv
\begin{pmatrix}
0 & 0 & \cdots & 0 & 1 & 0 & \cdots & 0 \\
0 & 0 & \cdots & 0 & 0 & 1 & \ddots & 0 \\
\vdots & \vdots & \ddots & \vdots & \vdots & \ddots & \ddots & \vdots \\
0 & 0 & \cdots & 0 & 0 & 0 & \cdots & 1 \\
-1 & 0 & \cdots & 0 & 0 & 0 & \cdots & 0 \\
0 & -1 & \ddots & 0 & 0 & 0 & \cdots & 0 \\
\vdots & \ddots & \ddots & \vdots & \vdots & \vdots & \ddots & \vdots \\
0 & 0 & \cdots & -1 & 0 & 0 & \cdots & 0
\end{pmatrix}. \quad (1.2.3)
$$

3. *Any skew-symmetric $(2n \times 2n)$ matrix A with* $\det A \neq 0$ *gives rise to the following symplectic form on $\mathbb{R}^{2n}$*

$$\omega(u, v) = u \cdot Av.$$

For instance the following matrices (Zehnder)

$$
A = \begin{pmatrix}
0 & 1 & 0 & -a_1 \\
-1 & 0 & 0 & -a_2 \\
0 & 0 & 0 & -a_3 \\
a_1 & a_2 & a_3 & 0
\end{pmatrix}
$$

where $(a_i)_{1 \leqslant i \leqslant 3}$ with $a_3 \neq 0$ are reals numbers.

Example 1.2

Let U be a n–dimensional vector space and $V = U \oplus U^*$. We define a symplectic form ω on V by:

$$\omega(u_1 \oplus u_1^*, u_2 \oplus u_2^*) = u_1^*(u_2) - u_2^*(u_1) \qquad \forall\, u = (u_1, u_1^*)\,,\; v = (u_2, u_2^*) \in V.$$

This shows that there are infinitely many symplectic forms on a vector space.

However, we prove that they are all "isomorphic".

We need:

Theorem 1.2 *(canonical basis)*

Let (V, ω) be a symplectic vector space. There exists a basis $\mathcal{E} = (e_1, \cdots, e_{2n})$ of V such that if $\mathcal{E} = (\varepsilon_1, \cdots, \varepsilon_{2n})$ is its dual basis, then

$$\omega = \sum_{j=1}^{n} \varepsilon_j \wedge \varepsilon_{j+n}.$$

Definition 1.2

The basis $\mathcal{E} = (\varepsilon_1, \cdots, \varepsilon_{2n})$ above is called a canonical basis (it is not unique).

Proof (Construction of a canonical basis):

Pick $e_1 \in V$, $e_1 \neq 0$. Then $\tilde{\omega}(e_1) \neq 0$ since $\tilde{\omega}$ is an isomorphism. Now pick $e_{n+1} = f_1 \in V$ such that $\tilde{\omega}(e_1)(f_1) = 1$. Let $V_1 = Span\{e_1, f_1\}$ be the subspace of V spanned by e_1 and f_1. Then $\dim V_1 = 2$ since if there exists any λ such that $e_1 = \lambda f_1$, $\tilde{\omega}(e_1)(f_1) = \omega(e_1, f_1) = 0$ since ω is antisymmetric.

Define

$$V_1^{\omega} = \{x \in V; \omega(x, v) = 0, \forall v \in V_1\}.$$

This is a subspace of V of dimension $2n - 2$. We have: $V_1 \cap V_1^{\omega} = \{0\}$. Indeed if $v \in V_1 \cap V_1^{\omega}$, $v = \lambda e_1 + \mu f_1$ for some reals λ and μ. But $\mu = \omega(v, e_1) = 0$ and $\lambda = \omega(v, f_1) = 0$; hence $v = 0$.

We want to show that

$$V = V_1 \oplus V_1^{\omega}.$$

Let $v \in V$. Set $\omega(v, e_1) = c$ and $\omega(v, f_1) = d$. Therefore

$$v = \underbrace{(-cf_1 + de_1)}_{\in V_1} + (v + cf_1 - de_1)$$

and $\quad \omega(v + cf_1 - de_1, e_1) = c - c = 0 \quad$ and $\quad \omega(v + cf_1 - de_1, f_1) = d - d = 0$
which implies that $v + cf_1 - de_1 \in V_1^\omega$.

Hence $V_1 \oplus V_1^\omega$ and on V_1 one has $\omega(e_1, f_1) = 1$.

Now we proceed the same way on V_1^ω.

Pick $e_2 \neq 0$ in V_1^ω and $f_2 = e_{n+2}$ with $\omega(e_2, f_2) = 1$ and look at $V_2 = \{e_2, f_2\}$ and V_2^ω; etc. We get

$$V = V_1 \oplus V_1^\omega = V_1 \oplus V_2 \oplus V_2^\omega = V_1 \oplus V_2 \oplus V_3 \oplus \cdots \oplus V_n.$$

Each V_j is 2–dimensional and then e_j, $f_j = e_{j+n}$ are so that $\omega(e_j, f_j) = 1$, for all $j = 1, 2, \cdots, n$; of course $\omega(e_i, f_j) = \omega(e_j, f_i) = 0$ for $i \neq j$ and $\omega(e_i, e_j) = \omega(f_i, f_j) = 0$.

Hence if $(\varepsilon_1, \cdots \varepsilon_{2n})$ is the standard dual basis of $(e_1, \cdots, e_{2n}, e_{n+1} = f_1, e_{n+2} = f_2, \cdots, e_{2n} = f_n)$, the symplectic form writes:

$$\omega = \sum_{j=1}^{n} \varepsilon_j \wedge \varepsilon_{j+n}.$$

$\square$

Exercise 1.1

1. *Find a canonical basis for the symplectic vector space* $(\mathbb{R}^4, \omega)$ *where* $\omega(u, v) = u \cdot Av$ *with*

$$A = \begin{pmatrix} 0 & 1 & 0 & -a_1 \\ -1 & 0 & 0 & -a_2 \\ 0 & 0 & 0 & -a_3 \\ a_1 & a_2 & a_3 & 0 \end{pmatrix} \qquad a_1, a_2, a_3 \ \text{are reals s.t } a_3 \neq 0.$$

2. *Same problem when*

$$A = \begin{pmatrix} 0 & 2 & -1 & 1 \\ -2 & 0 & -2 & -2 \\ 1 & 2 & 0 & 1 \\ -1 & 2 & -1 & 0 \end{pmatrix}.$$

1.3   Immediate consequence of Theorem 1.2

Corollary 1.1

Let (V, ω) be a symplectic vector space. There exists a hermitian structure (g, J) such that $\omega(u, v) = g(u, Jv)$.

Proof

Let $(e_1, \cdots, e_{2n}, e_{n+1} = f_1, e_{n+2} = f_2, \cdots, e_{2n} = f_n)$ be the canonical basis of (V, ω) constructed above. We define a linear map on V by

$$\begin{cases} J(e_j) & = & -e_{j+n} \\ J(e_{j+n}) & = & e_j. \end{cases}$$

We have $J^2 = -I$.

Now define $g : V \times V \longrightarrow \mathbb{R}$ by

$$g(e_i, e_j) = \delta_{ij} = \begin{cases} 1 & \text{if} & i = j \\ 0 & \text{if} & i \neq j. \end{cases} \tag{1.3.1}$$

This extends to a bilinear symmetric positive definite maps, i.e. a scalar product in V.

It is an easy exercise to prove that (g, J) is a hermitian structure and that $\omega(u, v) = g(u, Jv)$. $\qquad\square$

1.4   Another consequence of Theorem 1.2

Let (V, ω) and (V', ω') be two symplectic vector spaces of the same dimension $d = 2n$. Let $\mathscr{E} = (e_1, e_2, \cdots, e_{2n})$ and $\mathscr{E}' = (e'_1, e'_2, \cdots, e'_{2n})$ be canonical basis of (V, ω) and (V', ω').

The linear map

$$L : V \longrightarrow V'$$

such that

$$L(e_j) = e'_j$$

defines an isomorphism L^* between (antisymmetric) bilinear forms $\Lambda^2 V'$ onto $\Lambda^2 V$ by:

$$\bigl(L^*(b)\bigr)(u, v) = b\bigl(L(u), L(v)\bigr), \qquad \text{for all} \quad b \in \Lambda^2 V'.$$

Hence it takes ω' onto ω.

The linear map $L : V \longrightarrow V'$ is called a **symplectic isomorphism** and the symplectic forms ω and ω' are said to be isomorphic.

Proposition 1.2

All symplectic forms on vector spaces of the same dimension are isomorphic.

1.5 Compatible complex structures

Definition 1.3

Let (V, ω) be a symplectic vector space. A complex structure J on V is said to be **compatible** *with ω if*

1. $\omega(JX, JX') = \omega(X, X')$

2. *The bilinear form $(X, X') \longmapsto g_J(X, X') =: \omega(JX, X')$ is an inner product (i.e symmetric and positive definite).*

In that case we recover ω from g_J:

$$\omega(X, X') = g_J(X, JX').$$

Corollary 1.1 can be extended into:

Theorem 1.3

Any symplectic vector space (V, ω) admits a compatible complex structure. The set $\mathcal{J}(V, \omega)$ of all compatible complex structures is infinite and contractible.

Remark 1.1

The complex structure J in Corollary 1.1 is a compatible complex structure. The proof of the Theorem 1.3 provides a technique to construct more elements in $\mathcal{J}(V, \omega)$ in a uniform way.

Proof (of Theorem 1.3)

Let g be any inner product. Consider the dualities

$$\tilde{\omega} : \begin{array}{ccc} V & \longrightarrow & V^* \\ X & \longmapsto & \tilde{\omega}(X) \end{array} \qquad \text{and} \qquad \tilde{g} : \begin{array}{ccc} V & \longrightarrow & V^* \\ X & \longmapsto & \tilde{g}(X) \end{array}$$

such that for all $X, X' \in V$

$$\tilde{\omega}(X)(X') = \omega(X, X'); \qquad \tilde{g}(X)(X') = g(X, X')$$

and the linear isomorphism $A := \tilde{g}^{-1} \circ \tilde{\omega}$. We have

$$g(AX, X') = \omega(X, X').$$

Let ${}^t A$ be the transpose of A by the inner product g, we show that A is skew symmetric:

$$
\begin{aligned}
g({}^t AX, X') &= g(X, AX') = g(AX', X) \\
&= \omega(X', X) \\
&= -\omega(X, X') = -g(AX, X') \qquad \forall\, X \text{ and } X.
\end{aligned}
$$

The symmetric operator ${}^t A A$ is positive definite since

$$g({}^t AAX, X) = g(AX, AX) > 0 \qquad \forall\, X \neq 0.$$

Thus ${}^t A A$ is diagonalizable and has positive eigenvalues $\lambda_1, \cdots, \lambda_{2n}$. Hence it can be written ${}^t AA = \mathcal{B} \cdot diag(\lambda_1, \cdots, \lambda_{2n}) \cdot \mathcal{B}^{-1}$ for some matrix $\mathcal{B}$. We can define $\sqrt{{}^t AA}$ as

$$\mathcal{B} \cdot diag(\sqrt{\lambda_1}, \cdots, \sqrt{\lambda_{2n}}) \cdot \mathcal{B}^{-1}.$$

Denote $R = \sqrt{{}^t AA}$ and set $J = R^{-1}A$. Observe that R is a symmetric positive definite operator.

The decomposition $A = RJ$ is called the **polar decomposition**. We have $AJ = JA$. $\qquad\qquad\qquad\qquad\qquad\qquad\qquad\qquad\qquad$ □

Exercise 1.2

   Verify that

- $g(JX, JX') = g(X, X') \quad so \quad {}^t J \circ J = I,$

- $JR = RJ,$

- ${}^t J = -J \quad so \quad J^2 = -I.$

We now see that

$$
\begin{aligned}
\omega(JX, JX') &= g(AJX, JX') = g(JAX, JX') = g(AX, X') \\
&= \omega(X, X') \qquad\qquad\qquad\qquad\qquad\qquad (1.5.1)
\end{aligned}
$$

and $\omega(X, JX) = g(AX, JX) = g(-JAX, X) = g(RX, X) > 0 \quad \forall\, X \neq 0$.

We obtain the inner product

$$g_J(X, X') = \omega(X, JX') = g(AX, JX') = -g(JAX, X') = g(RX, X')$$

which is different from g. We recover ω for g_J as:

$$g_J(JX, X') = \omega(JX, JX') = \omega(X, X'). \tag{1.5.2}$$

Observe the construction of $J = J_g$ depend on a direct, explicit and canonical way of g. There are infinitely many of these $J = J_g$. Moreover if $J_1 = J_{g_1}$ and $J_2 = J_{g_2}$ then

$$J_t = J_{\left(tg_1 + (1-t)g_2\right)}$$

is a homotopy between J_1 and J_2. $\qquad\qquad\square$

1.6 The symplectic group

Let F be a linear map from a vector space U into a symplectic vector space (V, ω).

The pullback of ω on U is the 2-form $F^*\omega$ defined by:

$$(F^*\omega)(X, X') = \omega(FX, FX'). \tag{1.6.1}$$

If U is a symplectic vector space with the symplectic form ω_0, F is called a **symplectic map** if $F^*\omega = \omega_0$.

Exercise 1.3

Let (V, ω) be a symplectic vector space of dimension $2n$. Denote by $Sp(n, V)$ the set of all linear symplectic maps $F : (V, \omega) \longrightarrow (V, \omega)$. Show that $Sp(n, V)$ is a group.

Definition 1.4

*The group $Sp(n, V)$ is called the **symplectic group**.*

Remark 1.2

Let (V, ω) be a symplectic vector space, $B = (e_1, \cdots, e_n, f_1, \cdots, f_n)$ and $B = (e_1', \cdots, e_n', f_1', \cdots, f_n')$ be two canonical basis. The linear map T defined by $T(e_i) = e_i'$ and $T(f_i) = f_i'$ is a symplectic isomorphism. We say that its matrix is a symplectic matrix.

The elements of $Sp(n, V)$ are identified with symplectic matrices A, i.e $2n \times 2n$ matrices with real coefficients such that

$$\omega_0(AX, AX') \;=\; \omega_0(X, X') \tag{1.6.2}$$
$$\omega_0(AX, AX') \;=\; \langle AX, JAX' \rangle = \langle X, {}^tAJAX' \rangle = \langle X, JX' \rangle$$

where $J : \mathbb{R}^{2n} \longrightarrow \mathbb{R}^{2n}$ is the complex structure $J(u, v) = (v, -u)$. We have

$$\langle X, {}^tAJAX' - JX' \rangle = 0 \quad \forall \quad X \in \mathbb{R}^{2n}. \tag{1.6.3}$$

Hence
$$^tAJAX' = JX' \quad \forall \quad X', \qquad i.e \qquad J = {}^tAJA.$$

We conclude that:

Proposition 1.3
Any symplectic matrix A satisfies

$$A^{-1} = -J\,{}^tAJ.$$

Remark 1.3
If A is symplectic then ${}^tA$ is also symplectic, i.e

$$\omega({}^tAX, {}^tAX) = \omega(X, X), \qquad \forall X \tag{1.6.4}$$

or equivalently $AJ\,{}^tA = J$.

Chapter 2

Symplectic manifolds

The material in this chapter is developed in further details in the following references [Lib-Mar87], [Sil01], [Wei77]. The notions we need on calculus on manifolds are collected in Appendix A.

Definition 2.1

A symplectic form on a smooth manifold M is a differential form ω of degree 2 such that

1. $d\omega = 0$,

2. $\forall\, x \in M$, the 2-form ω_x on $T_x M$ is a symplectic form.

One also says that a symplectic form is a closed non-degenerate 2-form.

Remark 2.1

Since $T_x M$ is a symplectic vector space, its dimension must be even, i.e M must be even dimensional.

Exercise 2.1

If ω is a symplectic form on M, $\dim M = 2n$ for some integer n, then the non-degeneracy condition of ω is equivalent to the following condition:

$$\omega^n = \underbrace{\omega \wedge \cdots \wedge \omega}_{n \ \ times} \tag{2.0.1}$$

is everywhere non-zero.

The form ω^n defines an orientation of M.

One calls sometime ω^n the *"Liouville volume"* of M.

A **symplectic manifold** is a couple (M, ω) of a smooth $2n$-dimensional manifold equipped with a symplectic form.

13

2.1   Examples of symplectic manifolds

2.1.1   Euclidean spaces

On $\mathbb{R}^{2n}$ with the coordinates $(x_1, \cdots, x_n, y_1, \cdots, y_n)$

$$\omega_0 = dx_1 \wedge dy_1 + \cdots + dx_n \wedge dy_n \tag{2.1.1}$$

is a (canonical) symplectic form. Clearly $d\omega_0 = 0$ and ω_0 is non-degenerate.

The form defined by equation (2.1.1) is the local prototype of any symplectic form. Namely, we have the *Darboux* theorem:

Theorem 2.1

Let (M, ω) be a 2n-dimensional symplectic manifold. Each point $x \in M$ has an open neighborhood U which is the domain of a local chart $\varphi : U \longrightarrow \mathbb{R}^{2n}$ such that $\varphi(x) = 0$ and

$$\varphi^* \omega_0 = \omega|_U.$$

The theorem asserts that all symplectic manifold (locally) look alike. Therefore, there is no local invariants in symplectic geometry.

Let us now consider more examples of symplectic manifolds.

2.1.2   Tori

Let $T^{2n} = \mathbb{R}^{2n}/\mathbb{Z}^{2n}$ be the 2n-torus. Since the form ω_0 is invariant by translations it descends to the quotient T^{2n} and we denote it again by ω_0. Hence (T^{2n}, ω_0) is a symplectic manifold.

2.1.3   Oriented surfaces

Any oriented surface is a symplectic manifold, because the orientation form is itself symplectic.

For instance, on the sphere S^2, we define the symplectic (orientation form) this way:

let $x \in S^2$, $x = (x_1, x_2, x_3) \in \mathbb{R}^3$ with $\|x\| = 1$. So $X, X' \in T_x S^2$ are elements of $\mathbb{R}^3$ which are orthogonal to x. Define

$$\omega_x(X, X') = \det(x, X, X')$$

[Exercise: Show that it is a symplectic form.]

2.1.4 Product of symplectic manifolds

If (M_1, ω_1) and (M_2, ω_2) are symplectic manifolds, then for any $\lambda_1, \lambda_2 \in \mathbb{R}$ non-zero,

$$\omega_{\lambda_1, \lambda_2} = \lambda_1(p_1^* \omega_1) + \lambda_2(p_2^* \omega_2)$$

where $p_i : M_1 \times M_2 \longrightarrow M_i$ are the canonical projections on each factor, is a symplectic form. One often denotes it by $(\lambda_1 \omega_1) \oplus (\lambda_2 \omega_2)$.

2.1.5 Cotangent bundles

Let N be an n-dimensional manifold and $M = T^*N$ its cotangent bundle with projection $\pi : T^*N \longrightarrow N$. The **Liouville 1-form** λ_N on T^*N is defined as follows: let $a \in T^*N$ and $\xi \in T_a(T^*N)$; the differential of π at a is

$$T_a \pi : T_a T^*N \longrightarrow T_{\pi(a)} N;$$

denote by $x = \pi(a)$, i.e $a = (x, \theta_x)$ where $\theta_x \in T_x^* N$; hence $(T_a \pi)(\xi) \in T_x N$, therefore we can evaluate θ_x on it, and get:

$$\lambda_N(a)(\xi) = \theta_x \Big((T_a \pi)(\xi) \Big).$$

This is a canonical form, called the **Liouville 1-form**.

Exercise 2.2

Let $(x_1, \cdots, x_n)$ be local coordinates on $\mathcal{U} \subset N$ and $(x_1, \cdots, x_n, y_1, \cdots, y_n)$ the corresponding local coordinates on $T^*\mathcal{U}$, then

$$\lambda_N|_{T^*\mathcal{U}} = \sum_{i=1}^{n} y_i dx_i. \qquad (2.1.2)$$

As a consequence, we see that $\omega_N = d\lambda_N$ is a symplectic form on T^*N.

Exercise 2.3

    Let α be a 1-form on a smooth manifold N. View α as a section α : $N \longrightarrow T^*N$. Show that

$$\alpha^*(\lambda_N) = \alpha$$

where λ_N is the Liouville 1-form of N.

Exercise 2.4

    Let θ be a closed 2-form on N. Show that

$$\omega_{\theta,N} = d\lambda_N + \pi^*\theta$$

is a symplectic form on $M = T^*N$.

Sometimes this symplectic form is called the magnetic symplectic form.

2.2   The cohomology class of a symplectic form (see Appendix A)

Let (M, ω) be a symplectic manifold of dimension $2n$, with $\partial M = \emptyset$. Since ω is closed, it determines a cohomology class $[\omega] \in H^2(M, \mathbb{R})$.

    If M is compact, then $[\omega^n] \in H^{2n}(M, \mathbb{R})$ is non-zero, since

$$\int_M \omega^n \neq 0. \qquad (2.2.1)$$

Recall that ω^n is a volume form and M is oriented.

    In fact, the cohomology classes $[\omega] \in H^2(M, \mathbb{R})$ and $[\omega^p] \in H^{2p}(M, \mathbb{R})$, $1 \leqslant p \leqslant n$, are all non-zero. Indeed if $\omega^p = d\theta$,

$$\omega^n = \omega^p \wedge \omega^{n-p} = d(\theta \wedge \omega^{n-p}) \qquad (2.2.2)$$

and by Stokes theorem

$$\int_M \omega^n = \int_{\partial M} \theta \wedge \omega^{n-p} = 0$$

contradicting (2.2.1).

    Therefore no symplectic form on a compact manifold without boundary can be exact.

For instance the symplectic form ω_0 on T^{2n} coming from the canonical symplectic form ω_0 of $\mathbb{R}^{2n}$ is not exact to the contrary of ω_0 (on $\mathbb{R}^{2n}$):

$$\omega_0 = d\theta \qquad\qquad (2.2.3)$$

$$\theta = \sum_{i=1}^{n} y_i dx_i \qquad \text{or}$$

$$\theta = \pm\frac{1}{2}\sum_{i=1}^{n}\Big(x_i dy_i - y_i dx_i\Big).$$

We also conclude that the spheres $S^{2n}, n > 1$ cannot be symplectic since $H^2(S^{2n}, \mathbb{R}) = 0$ for $n > 1$.

The cohomology class $[\omega]$ is the first example of a (global) symplectic invariant.

2.3   Moser path method

In [Mos65], Moser proved the following:

Theorem 2.2

Let M be a compact manifold equipped with a smooth family of symplectic forms ω_t, $0 \leqslant t \leqslant 1$, which represent the same cohomology class $[\omega_t] \in H^2(M, \mathbb{R})$. Then there exists a smooth family of diffeomorphisms (an isotopy) $\varphi_t : M \longrightarrow M$ such that

$$\varphi_t^* \omega_t = \omega_0 \qquad and \qquad \varphi_0 = id.$$

Proof

The hypothesis means that $\left[\frac{\partial}{\partial t}\omega_t\right] = 0$. By the Hodge decomposition theorem, there exists a smooth family β_t of 1-forms such that

$$\frac{\partial}{\partial t}\omega_t = -d\beta_t.$$

Let X_t be the (smooth) family of vectors fields on M such that $i_{X_t}\omega_t = \beta_t$. Then

$$di_{X_t}\omega_t = d\beta_t = -\frac{\partial}{\partial t}\omega_t.$$

Hence

$$L_{X_t}\omega_t = (i_{X_t}d\omega_t) + di_{X_t}\omega_t = -\frac{\partial}{\partial t}\omega_t$$

or

$$L_{X_t}\omega_t + \frac{\partial}{\partial t}\omega_t = 0.$$

If φ_t is the family of diffeomorphisms obtained by integrating the differential equation

$$\frac{d}{dt}\varphi_t(x) = X_t\Big(\varphi_t(x)\Big), \qquad \varphi_0(x) = x$$

we have

$$\frac{d}{dt}\big(\varphi_t^*\omega_t\big) = \varphi_t^*\left(L_{X_t}\omega_t + \frac{\partial}{\partial t}\omega_t\right) = 0$$

which means that $\varphi_t^*\omega_t = \omega_0$.   □

Weinstein observed that this method yields a proof for *Darboux* theorem. This is why the Darboux theorem above is also called Darboux-Weinstein theorem.

2.3.1   Proof of Darboux-Weinstein Theorem

Let $\varphi : U \longrightarrow \mathbb{R}^{2n}$ be a coordinate chart around $x \in M$ such that $\varphi(x) = 0$ and let $V = \varphi(U) \subseteq \mathbb{R}^{2n}$. On V, consider the constant symplectic forms:

$$\Omega_0 \quad = \quad \big(\varphi^{-1}\big)^*(\omega|_U)(0) \qquad\qquad (2.3.1)$$

and

$$\omega_0 \quad = \quad \sum_{i=1}^{n} dx_i \wedge dy_i,$$

the canonical symplectic form on $\mathbb{R}^{2n}$.

There exists a linear map $\rho : \mathbb{R}^{2n} \longrightarrow \mathbb{R}^{2n}$ such that $\rho^*\Omega_0 = \omega_0$ (existence of the canonical basis). The symplectic form

$$\Omega = \rho^*\big(\varphi^{-1}\big)^*\omega|_U = \big(\varphi^{-1} \circ \rho\big)^*(\omega|_U)$$

and ω_0 coincide at $0 \in \mathbb{R}^{2n}$. Hence there is a neighborhood V_1 of 0, $V_1 \subset V$ such that $\omega_t = t\Omega + (1 - t)\omega_0$ is non-degenerate (the condition of non-degeneracy is an open condition).

We may assume that V_1 is star-like so that by Poincaré lemma $\Omega = d\beta$, with support of β in V_1. Therefore $\frac{\partial}{\partial t}\omega_t = d\big(\beta - \alpha_0\big)$ where $\alpha_0 = \sum_i x_i dy_i$.

We can modify β by adding an exact 1-form $d\rho$ and get $\beta' = \beta + d\rho$ so that $\gamma = \beta' - \alpha_0$ vanishes at 0. We still have $\frac{\partial}{\partial t}\omega_t = d\gamma$.

Let X_t be the family of vector fields on V_1 defined by

$$i_{X_t}\omega_t = -\gamma. \tag{2.3.2}$$

Clearly $X_t(0) = 0$. The flow ψ_t of X_t fixes 0 and hence sends a small neighborhood V_0 of 0 into $V_2 \subset V_1$.

The equation (2.3.2) implies that

$$L_{X_t}\omega_t = i_{X_t}d\omega_t + di_{X_t}\omega_t \quad = \quad -d\gamma = -\frac{\partial}{\partial t}\omega_t \tag{2.3.3}$$

or

$$L_{X_t}\omega_t + \frac{\partial}{\partial t}\omega_t \quad = \quad 0.$$

By (2.2.1) this means that

$$\psi_t^*\omega_t = \omega_0$$

on V_0. Therefore

$$\psi_1^*\Omega = \omega_0$$

on V_0. The required chart is $(\varphi^{-1} \circ \rho \circ \psi_1)^{-1} = \psi_1^{-1} \circ \rho^{-1} \circ \varphi$. □

Remark 2.2

Let $B_r^{2n} = \{(x,y) \in \mathbb{R}^{2n} \text{ s.t } x_1^2 + \cdots + x_n^2 + y_1^2 + \cdots + y_n^2 < r\} \subset V_0$. We just constructed an embedding

$$\mu : B_r^{2n} \longrightarrow M$$

such that $\mu^*\omega = \omega_0$ where ω_0 is the restriction of the standard form ω_0 to the ball B_r^{2n}.

Gromov [Gro85] defined the following number:

$$G(M,\omega) \quad = \quad \sup_r \left\{ r \text{ s.t } \exists \text{ symplectic embedding } \mu : B_r^{2n} \longrightarrow M \right\}$$

$$\in \quad \mathbb{R} \cup \{+\infty\}.$$

He proved that this number is a symplectic invariant which is known as the **Gromov capacity**.

In general a capacity is defined as follows:

Definition 2.2

A *symplectic capacity is a function c defined on the category $\mathcal{S}_{2n}$ of symplectic manifolds of dimension $2n$*

$$c : \mathcal{S}_{2n} \longrightarrow \mathbb{R} \cup \{\infty\}$$

such that

(i) *if there is a symplectic embedding $(M_1, \omega_1) \hookrightarrow (M_2, \omega_2)$ then $c(M_1, \omega_1) \leqslant c(M_2, \omega_2)$.*

(ii) *$c(M, \lambda\omega) = |\lambda| c(\omega)$ where λ is any non-zero number.*

(iii) *$c(Z_r, \omega_0|_{Z_r}) = c(B_r) = 2\pi r$ where*

$$Z_r = \left\{ (x_1, \cdots, x_n, y_1, \cdots, y_n)/(x_1, y_1) \in B_r^2 \right\}.$$

Exercise 2.5

Show that $G(M, \omega)$ is a symplectic capacity.

Property (i) says that a symplectic capacity is a symplectic invariant.

2.4   Symplectomorphisms

Let (M_1, ω_1) and (M_2, ω_2) be two symplectic manifolds. A smooth diffeomorphism $\varphi : M_1 \longrightarrow M_2$ is said to be a **symplectomorphism** or a **symplectic diffeomorphism** if

$$\varphi^* \omega_2 = \omega_1.$$

We say that (M_1, ω_1) and (M_2, ω_2) are **symplectomorphic** if there exists a symplectomorphism $\varphi : M_1 \longrightarrow M_2$.

Let $Symp(M, \omega)$ be the set of all symplectomorphisms $\varphi : M \longrightarrow M$ of (M, ω) into itself.

Exercise 2.6

1. Let (M, ω) be a symplectic manifold. Show that $Symp(M, \omega)$ is a group.

2. *Given two symplectic forms ω_1 and ω_2 on a smooth manifold M, we write*

$$\omega_1 \sim \omega_2 \Longleftrightarrow \exists\, \varphi \quad \text{a diffeomorphism such that} \quad \varphi^* \omega_1 = \omega_2.$$

Show that "$\sim$" is an equivalence relation.

Example 2.1

1. *Let $v = (a, b) \in \mathbb{R}^{2n}$.*

$$
\begin{array}{rcl}
T_v: \quad \mathbb{R}^{2n} & \longrightarrow & \mathbb{R}^{2n} \\
(x, y) & \longmapsto & (x + a, y + b)
\end{array}
$$

is a symplectomorphism of $(\mathbb{R}^{2n}, \omega_0)$. This symplectomorphism descends to T^{2n}.

2. *For $a = (a_1, \cdots, a_n)$, $a_i \neq 0$.*

$$\chi_a(x_1, \cdots, x_n, y_1, \cdots, y_n) = (a_1 x_1, \cdots, a_n x_n, \frac{1}{a_1} y_1, \cdots, \frac{1}{a_n} y_n)$$

is a symplectomorphism of $(\mathbb{R}^{2n}, \omega_0)$.

2.4.1 A general method for constructing symplectomorphisms

Let θ be a closed 1-form with compact support on a symplectic manifold (M, Ω).

Since

$$
\begin{array}{rcl}
\tilde{\omega}: \quad TM & \longrightarrow & T^*M \\
X & \longmapsto & \tilde{\omega}(X) := i_X \omega
\end{array}
$$

is an isomorphism, there exists a unique vector field with compact support such that

$$i_X \omega = \theta.$$

We have

$$L_X \omega = d i_X \omega + i_X d\omega = d\theta = 0.$$

The flow $\varphi_t : M \longrightarrow M$ of X, i.e the family of diffeomorphisms $(\varphi_t)_t$ with $\varphi_0 = id$ and

$$\frac{d}{dt}\varphi_t(x) = X_{\varphi_t(x)}$$

satisfies
$$\varphi_t^* \omega = \omega.$$

Definition 2.3

The support of a diffeomorphism $\phi : M \longrightarrow M$ is the closure of the set $\{x \in M,\ \phi(x) \neq x\}$.

One denote this set by $\mathrm{Supp}(\phi)$.

The sets $\mathrm{Diff}_c(M)$, $\mathrm{Symp}_c(M, \omega)$ of smooth diffeomorphisms and symplectomorphisms with compact support are subgroup of $\mathrm{Diff}(M)$ and $\mathrm{Symp}_c(M)$. We will endow them with the $\mathcal{C}^\infty$-compact open topology.

For $r > 0$, we denote by B_r the open ball $\{x \in \mathbb{R}^p\ \|x\| < r\}$ in $\mathbb{R}^p$. We have the following

Proposition 2.1

Let $v \in \mathbb{R}^{2n}$ with $\|v\| = r$. There exists $\phi \in Symp(\mathbb{R}^{2n}, \omega_0)$ whose support is inside B_{5r} and which is equal to the translation T_v on B_r.

Proof

The translation T_v, where $v = (a, b)$ is generated by the constant vector field
$$X = a_1 \frac{\partial}{\partial x_1} + \cdots + a_n \frac{\partial}{\partial x_n} + b_1 \frac{\partial}{\partial y_1} + \cdots + b_n \frac{\partial}{\partial y_n}.$$

The corresponding 1-form (via the symplectic form) $\sum_i dx_i \wedge dy_i$ is

$$i_X \omega = \sum_{i=1}^{n} a_i dy_i - b_i dx_i = dH$$

where $H = \sum_i (a_i y_i - b_i x_i)$. Choose a smooth function $\lambda : \mathbb{R}^{2n} \longrightarrow \mathbb{R}$ such that

$$\lambda = \begin{cases} 1 & \text{on} & B_{4r} \\ 0 & \text{outside of} & B_{5r}. \end{cases}$$

Then the vector field $\widetilde{X}$ corresponding to the 1-form $d(\lambda H)$, i.e

$$i_{\widetilde{X}} \omega = d(\lambda H) \tag{2.4.1}$$

generates a 1-parameter group of symplectomorphisms ϕ_t supported in B_{5r} and equal to T_v on B_r. $\qquad\qquad\qquad\qquad\qquad\qquad\qquad\qquad\qquad\qquad$ $\square$

Remark 2.3

The diffeomorphism $\varphi = \varphi_1$ constructed maps $0_{\mathbb{R}^n}$ to v.

Corollary 2.1

Let $U \subseteq M$ be open subset of a symplectic manifold (M, ω), which is a domain of a symplectic chart $U \longrightarrow \mathbb{R}^{2n}$. For any open subset $V \subset \overline{V} \subset U$ and any points $x, y \in V$, there exists a symplectic diffeomorphism with support in U and which takes x to y.

From the Corollary 2.1, it is easy to deduce the following important result [Boo69]:

Theorem 2.3 *(Boothby)*

Let (M, ω) be a connected symplectic manifold. Then $Symp(M, \omega)$ acts p-transitively on M, i.e given two sets $(x_1, \cdots, x_p)$ and $(y_1, \cdots, y_p)$ of distinct points on M, there exists $\varphi \in Symp(M, \omega)$ such that

$$\varphi(x_i) = y_i \qquad\qquad \forall \, i = 1, \cdots, p.$$

Proof

Choose a smooth path $\gamma(t)$ from x_1 to y_1 which does not pass through x_i and y_i for all $i > 1$. Consider a subdivision

$$0 = t_0 < t_1 < \cdots < t_N = 1$$

fine enough so that each pair $\big(\gamma(t_i), \gamma(t_{i+1})\big)$ satisfies the condition of Corollary 2.1: there exists small open set U_i and symplectic diffeomorphism ψ_i such that $\operatorname{supp}(\psi_i) \subset U_i$ and

$$\psi_i\big(\gamma(t_i)\big) = \gamma(t_{i+1}).$$

Now define

$$\phi^1 = \psi_N \circ \cdots \circ \psi_2 \circ \psi_1.$$

Clearly $\phi^1(x_1) = y_1$ and $\operatorname{supp}(\phi^1) \subset \bigcup_{i=1} U_i \subset K_1$ for some compact subset K_1.

We may arrange that K_1 does not contain the points $x_2, \cdots, x_p, y_2, \cdots, y_p$.

Now pick another path γ_2 disjoint of K_1 joining x_2 to y_2 and repeat the procedure to construct ϕ^2.

Similarly, we construct ϕ^j taking x_j to y_j for $j = 3, \cdots, p$, with disjoint support contained in K_j.

Finally, the desired diffeomorphism is the diffeomorphism ϕ which restricts to ϕ^j on K_j for all $j = 1, \cdots, p$. $\qquad\square$

Exercise 2.7

Let $\varphi : N_1 \longrightarrow N_2$ be a diffeomorphism between two n-dimensional manifolds N_1 and N_2. Construct a diffeomorphism $\tilde{\varphi} : T^*N_1 \longrightarrow T^*N_2$ such that

$$
\begin{array}{ccc}
T^*N_1 & \overset{\tilde{\varphi}}{\longrightarrow} & T^*N_2 \\
\pi_1 \downarrow & & \downarrow \pi_2 \\
N_1 & \overset{\varphi}{\longrightarrow} & N_2
\end{array}
$$

commutes and $\tilde{\varphi}^*(\lambda_{N_2}) = \lambda_{N_1}$. (In particular $\tilde{\varphi}^*$ is a symplectomorphism of $(T^*N_1, d\lambda_{N_1})$ into $(T^*N_2, d\lambda_{N_2})$).

Exercise 2.8

Show that if $\theta = d\beta$ is an exact 2-form and $\omega_{\theta,N} = d\lambda_N + \pi^*\theta$, there exists a symplectomorphism

$$
\varphi : (T^*N, \omega_{\theta,N}) \longrightarrow (T^*N, d\lambda_N).
$$

The group $Symp(M,\omega)$ determines the symplectic geometry of (M,ω).

Boothby theorem implies that any connected symplectic manifold (M,ω) can be viewed as a homogeneous space:

$$
(M,\omega) = \mathrm{Diff}(M,\omega)/_{\mathrm{Diff}(M,\omega,x_0)} \tag{2.4.2}
$$

where $\mathrm{Diff}(M,\omega,x_0)$ is the isotopy subgroup of some point x_0.

A much deeper result of [Ban86, Ban88, Ban97] asserts:

Theorem 2.4 *(Banyaga)*

Let (M_i, ω_i), $i = 1, 2$ be two symplectic manifolds. Suppose that there exists a group isomorphism

$$\Phi : Symp(M_1, \omega_1) \longrightarrow Symp(M_2, \omega_2),$$

then there is a diffeomorphism $\varphi : M_1 \longrightarrow M_2$ such that

$$\begin{cases} \varphi^* \omega_2 = \lambda \omega_1 & \text{for some constant } \lambda \\ \Phi(h) = \varphi \circ h \circ \varphi^{-1} & \forall\, h \in Symp(M_1, \omega_1). \end{cases}$$

2.4.2   The Calabi homomorphism [Ban78], [Ban97]

Let $\text{Diff}_\theta(M)$ be the group of all $\mathcal{C}^\infty$ diffeomorphisms φ of a smooth manifold M with compact support and which preserve a closed p-form θ on M i.e $\varphi^* \theta = \theta$.

An isotopy in $\text{Diff}_\theta(M)$ is a smooth map $\Phi : M \times [0, 1] \longrightarrow M$ such that if $\varphi_t : M \longrightarrow M$ denotes the map $\varphi_t(x) = \Phi(x, t)$ then $\varphi_t \in \text{Diff}_\theta(M)$.

An element $\varphi \in \text{Diff}_\theta(M)$ is said to be isotopic to the identity if there exists an isotopy Φ in $\text{Diff}_\theta(M)$ such that $\varphi_1 = \varphi$.

We denote by $\text{IDiff}_\theta(M)$ the set of all isotopies in $\text{Diff}_\theta(M)$ and by $\text{Diff}_\theta(M)_0$ the set of all $\varphi \in \text{Diff}_\theta(M)$ which are isotopic to the identity. Clearly, both $\text{IDiff}_\theta(M)$ and $\text{Diff}_\theta(M)_0$ are groups.

On $\text{IDiff}_\theta(M)$ we put the following equivalence relation:

$$\Phi = (\varphi, t) \simeq \Psi = (\psi, t) \Longleftrightarrow \varphi_1 = \psi_1$$

and there exists a smooth map $u : M \times [0, 1] \times [0, 1] \longrightarrow M$ such that if $u_{(s,t)} : M \longrightarrow M$ stands for $u_{(s,t)}(x) = u(x, s, t)$ then

$$\begin{aligned} u_{(s,t)} &\in \text{Diff}_\theta(M) \\ u_{(0,t)} &= \varphi_t & u_{(1,t)} &= \psi_t \\ u_{(s,0)} &= \text{Idendity} & u_{(s,1)} &= \varphi_1 = \psi_1. \end{aligned}$$

We denote by $\widetilde{\text{Diff}_\theta}(M)$ the set of equivalence classes, i.e $\widetilde{\text{Diff}_\theta}(M) = \text{IDiff}_\theta(M)/_\simeq$

Exercise 2.9

Show that $\widetilde{\text{Diff}_\theta}(M)$ is a group and there is a natural map $\Pi : \widetilde{\text{Diff}_\theta}(M) \longrightarrow \text{Diff}_\theta(M)_0$ which is a surjective homomorphism.

Remark 2.4

If $\mathrm{Diff}_\theta(M)$, endowed with the C^∞-compact topology is locally connected by arcs then the group $\widetilde{\mathrm{Diff}}_\theta(M)$ is the universal cover of the identity component $\mathrm{Diff}_\theta(M)_0$ of $\mathrm{Diff}_\theta(M)$.

Exercise 2.10

Show that for each $\Phi = (\varphi_t) \in \mathrm{IDiff}_\theta(M)_0$ the cohomology class

$$\rho(\varphi_t) = [\rho] \in H^{p-1}(M, \mathbb{R})$$

of

$$\rho = \int_0^t \varphi_s^* \left(i_{X_s} d\theta \right) ds$$

depends only on the equivalence class $[\Phi] \in \widetilde{\mathrm{Diff}}_\theta(M)$ of Φ.
Hence it defines a map

$$\tilde{S} : \widetilde{\mathrm{Diff}}_\theta(M) \longrightarrow H^{p-1}(M, \mathbb{R}).$$

Exercise 2.11

Show that $\tilde{S}$ is a surjective group homomorphism (called the **Calabi homomorphism**).
Let

$$\Gamma = \tilde{S}\left(\ker \Pi : \widetilde{\mathrm{Diff}}_\theta(M) \longrightarrow \mathrm{Diff}_\theta(M)_0 \right).$$

Then $\tilde{S}$ induces also a surjective homomorphism

$$S : \mathrm{Diff}_\theta(M) \longrightarrow H^{p-1}(M, \mathbb{R})/_\Gamma.$$

Suppose now that θ is a symplectic form ω.

Proposition 2.2

The group $\mathrm{Diff}_\omega(M)$ with the C^∞-compact open topology is locally connected by arcs.

Proof

This follows from the existence of the Weinstein chart which will be established in the next section. $\qquad\square$

Hence $\widetilde{\mathrm{Diff}_\omega}(M)_0$ is the universal cover of the identity component $\mathrm{Diff}_\omega(M)_0$ of $\mathrm{Diff}_\omega(M)$.

The homomorphisms above

$$\widetilde{S} : \widetilde{\mathrm{Diff}_\omega}(M) \longrightarrow H^1(M,\mathbb{R}) \quad \text{and} \quad S : \mathrm{Diff}_\omega(M)_0 \longrightarrow H^1(M,\mathbb{R})/\Gamma$$

were first considered by Calabi [Cal70].

We have the following fundamental result is symplectic topology [Ban78].

Theorem 2.5 *(Banyaga)*
If M is compact, then the kernel of S, $\ker S$ is a simple group (i.e it does not admit a non-trivial normal subgroup).

In particular $\ker S$ is equal to its commutator subgroup: $[\ker S, \ker S]$.

2.5 Lagrangian submanifolds

Definition 2.4
Let (M,ω) be a $2n$-dimensional symplectic manifold. A n-dimensional submanifold $L \subset M$ is called a **Lagrangian submanifold** *if $j^*\omega = 0$ where $j : L \hookrightarrow M$ is the embedding of L into M. Such an embedding is called a* **Lagrangian embedding**.

2.5.1 Examples

1. In $\mathbb{R}^{2n}$ with coordinates (x,y); $x,y \in \mathbb{R}^n$ and symplectic form

$$\omega_0 = \sum_{i=1}^{n} dx_i \wedge dy_i,$$

   the following sets $L_1 = \{(x,y) \text{ s.t } y = 0\}$, $L_1' = \{(x,y) \text{ s.t } x = 0\}$, $L_2 = \{(x,y) \text{ s.t } x = y\}$ are all Lagrangian submanifolds.

2. The set L_1 above generalizes as follow. Let N be any n-dimensional manifold and $(M,\omega) = (T^*N, d\lambda)$. The zero section $\sigma_0 : N \longrightarrow M = T^*N$ assigns to each $x \in N$, $(x, 0_x)$ where $0_x \in T_x^*N$ is the zero linear form. In local coordinates (x,y), it looks like example L_1 above.

   Hence identifying $\sigma_0(N) \simeq N$, we see that any $\mathcal{C}^\infty$ manifold N is a Lagrangian submanifold of T^*N.

3. Let α be a 1-form on N, then $\alpha : N \longrightarrow T^*N$ embeds the graph of α into T^*N. Since $\alpha^*\lambda_N = \alpha$

$$d\alpha = d\alpha^*\lambda_N = \alpha^*(d\lambda_N).$$

Therefore $\alpha^*(d\lambda_N) = 0 \Longleftrightarrow d\alpha = 0$.

Hence the graph Γ_α of a 1-form α is a Lagrangian submanifold if and only if α is closed.

In particular for any smooth function $f : N \longrightarrow \mathbb{R}$,

$$\Gamma_{df} \subset T^*N$$

is a Lagrangian submanifold.

4. Any smooth embedded curve in an oriented surface (M, ω) is a Lagrangian submanifold. The curve is a 1-dimensional manifold and the restriction of a 2-form on it is automatically zero.

5. The graph Γ_φ of a symplectomorphism $\varphi : M \longrightarrow M$ of a symplectic manifold (M, ω)

$$j : M \longrightarrow M \times M, \ x \longmapsto (x, \varphi(x))$$

is a Lagrangian submanifold of $\big(M \times M, \omega \oplus (-\omega)\big)$. Indeed

$$j^*\big(\omega \oplus (-\omega)\big) = j^* p_1^* \omega - j^* p_2^* \omega = (p_1 \circ j)^* \omega - (p_2 \circ j)^* \omega = \omega - \varphi^* \omega = 0.$$

These examples show that smooth manifold, smooth functions on manifold, closed one form, symplectic diffeomorphisms, $\cdots$, are Lagrangians submanifolds in some symplectic manifold. This made Weinstein proclaim

Weinstein Creed: "Everything is a Lagrangian submanifold".

We will see more examples of Lagrangian submanifolds and some of their properties later in these lectures.

Remark 2.5

Let $j : L \longrightarrow (T^*N, d\lambda_N)$ be a Lagrangian submanifold in T^*N, the 1-form $\lambda = j^*(\lambda_N)$ is closed since

$$d\lambda = d(j^*\lambda_N) = j^*(d\lambda_N) = 0.$$

Hence λ determines a cohomology class $[\lambda] \in H^1(L, \mathbb{R})$.

For instance let $j : L \longrightarrow T^*\mathbb{R}^n \simeq \mathbb{R}^{2n}$ be a Lagrangian submanifold and let γ be a cycle in L representing a homology class $\alpha \in H_1(L)$, then

$$\langle [\lambda], \alpha \rangle = \int_\gamma \lambda_N = \int_\Sigma d\lambda_N$$

where Σ is a 2-chain (a surface in $\mathbb{R}^{2n}$ bounded by γ). One easily shows that $\langle [\lambda], \alpha \rangle$ does not depend on the chain Σ and can be interpreted as "an area".

Neighborhood of a Lagrangian submanifold

The following result is due to Kostant and Weinstein [Wei71].

Theorem 2.6

Let $L^n \hookrightarrow (M^{2n}, \omega)$ be a compact Lagrangian submanifold of a symplectic manifold (M^{2n}, ω). There exist a neighborhood $\mathcal{U}(L)$ of L in M, a neighborhood $\mathcal{V}(L_0) \subset T^*L$ of the zero section and a symplectomorphism $\Phi : \mathcal{U}(L) \longrightarrow \mathcal{V}(L_0)$ such that

$$\Phi^*(d\lambda_L) = \omega \qquad and \qquad \Phi|_L = id$$

where λ_L is the canonical (Liouville) 1-form on T^*L.

Example 2.2

Let (M, ω) be a symplectic manifold and consider $\triangle = \{(x, x) \quad x \in M\}$ the diagonal in $M \times M$. It is a Lagrangian submanifold of $(M \times M, \omega \oplus (-\omega))$ as the graph of the identity. The theorem implies that a neighborhood $\mathcal{N}(\triangle)$ of $\triangle$ is symplectomorphic to a neighborhood $\mathcal{V}(M_0)$ of the zero section in T^*M.

Suppose now that $\varphi \in \text{Symp}(M, \omega)$ is sufficiently $\mathcal{C}^1$-close to the identity, then

$$L = \Phi(graph\varphi)$$

is a Lagrangian submanifold of $\mathcal{U}(\triangle)$, which is $\mathcal{C}^1$-close to the canonical embedding of the zero section.

Hence L is diffeomorphic to the graph of a 1-form on M, denoted $W(\varphi)$, which is closed since it is Lagrangian, and called **Weinstein form of** φ [Ban80].

The Weinstein chart

Let $V \subset \text{Symp}(M, \omega)$ be an open neighborhood of the identity (in the $\mathcal{C}^1$-topology) small enough so that the graph of any $\varphi \in V$ is contained in $\mathcal{U}(\Delta)$.

For any $\varphi \in V$ the Weinstein form $W(\varphi)$ belongs to a neighborhood W of zero in the space of closed 1-form on M.

The correspondance

$$\begin{aligned} \mathcal{W}: \quad V &\longrightarrow W \\ \varphi &\longmapsto W(\varphi) \end{aligned} \qquad (2.5.1)$$

is called the Weinstein chart.

The zeros of $W(\varphi)$ are in 1-1 correspondence with fixed points of φ (intersections of the graph of φ and the diagonal which is the graph of the identity).

If $W(\varphi)$ is exact, i.e there exists $f \in \mathcal{C}^\infty(M)$ such that $W(\varphi) = df$, then the zeros of $W(\varphi)$ coincide with critical points of f.

Since on a compact manifold, every smooth function has at least one critical point, we get:

Theorem 2.7 *[Wei71]*

Let M be a compact simply connected symplectic manifold and $\varphi \in Symp(M, \omega)$ which is $\mathcal{C}^1$-close to the identity, then φ has at least one fixed point.

2.6   Compatible almost complex structures

In Section 1.5, we studied compatible structures in a symplectic vector space.

In this section, we carry out the construction on a symplectic manifold.

An **almost complex structure** on a smooth manifold M is a bundle map

$$J : TM \longrightarrow TM$$

such that for all $x \in M$

$$J_x : T_x M \longrightarrow T_x M$$

satisfies $J_x^2 = -I_{T_x M}$.

Let (M, ω) be a symplectic manifold. Choose a Riemannian metric g on M.

Apply the construction in Theorem 1.3 to each g_x and ω_x on $T_x M$. We get $J_x \in \mathcal{J}(T_x M, \omega_x)$ and a metric

$$g_{J_x}(u, v) = \omega_x(u, J_x v). \tag{2.6.1}$$

Since the construction is canonical we get an almost complex structure J such that

$$\omega(X, X') = \omega(JX, JX') \tag{2.6.2}$$

and

$$g_J : (X, X') \longmapsto \omega(X, JX')$$

is a Riemannian metric.

The almost complex structure J is said to be compatible with ω. Moreover, we recover ω from g_J by

$$\omega(X, X') = g_J(JX, X').$$

We obtain the following:

Theorem 2.8

Let ω be a symplectic form on a smooth manifold M. The set $\mathcal{J}(M, \omega)$ of almost complex structures compatible with ω is infinite and contractible.

2.7   Almost Kaehler structures

We just saw that a symplectic manifold always carries an almost complex structure.

Consider now a manifold M carrying an almost complex structure J. Pick any Riemannian metric g_0 and consider the Riemannian metric g defined by:

$$g(X, Y) = g_0(X, Y) + g_0(JX, JY). \tag{2.7.1}$$

Then g is hermitian. By Exercise 1.1 the 2-form

$$\omega(X, Y) = g(X, JY)$$

is non-degenerate.

However it may not be closed (and hence not a symplectic form).

For instance on S^6 there is an almost complex structure (Calabi-Eckman) but we know that S^6 has no symplectic form.

If $d\omega = 0$ and J is a complex structure (i.e an integrable almost complex structure), the couple (g, J) is called a **Kaehler structure** and the form ω a **Kaehler form**.

For long time, it was believe that any symplectic form is Kaehler until Thurston found a simple conter example [Thu76].

Chapter 3

Hamiltonian systems and Poisson algebra

3.1 Hamiltonian systems

On a symplectic manifold (M, ω), a $\mathcal{C}^\infty$ function $f : M \longrightarrow \mathbb{R}$ determines uniquely a vector field X_f, called the **Hamiltonian vector field** with the Hamiltonian f by the equation

$$i_{X_f}\omega = df$$

or

$$X_f = \tilde{\omega}^{-1}(df)$$

where $\tilde{\omega} : TM \longrightarrow T^*M$ is the isomorphism $\tilde{\omega}(X) = i_X\omega$.

An immediate property of this vector field is that

$$X_f \cdot f = 0,$$

i.e the function is invariant under the flow of X_f (Physicists call this the principle of conservation of kinetic energy). Indeed

$$X_f \cdot f = df(X_f) = (i_{X_f}\omega)(X_f) = \omega(X_f, X_f) = 0$$

since ω is skew symmetric.

Let M_f be a regular "**energy**" surface, $M_f = f^{-1}(c)$ where c is a regular value of f.
Then the vector field X_f is tangent to M_f.

The second important property of X_f is that its (local) flow φ_t^f preserves the symplectic form ω:

Indeed

$$
\begin{aligned}
L_{X_f}\omega &= di_{X_f}\omega + i_{X_f}d\omega \qquad\qquad (3.1.1)\\
&= d(df) + 0\\
&= 0.
\end{aligned}
$$

By (3.1.1), the local flow of X_f preserves ω. If M is compact or the function $f : M \longrightarrow \mathbb{R}$ has compact support, then X_f is integrable and generates a global flow $\{\varphi_t^f\}$ where each φ_t^f is a symplectomorphism.

Exercise 3.1
Let $M = \mathbb{R}^{2n}$ *with coordinates* $(x,y) = (x_1,\cdots,x_n,y_1,\cdots,y_n)$ *and the symplectic form*
$\omega = \sum\limits_{i=1}^{n} dx_i \wedge dy_i$. *Let* $H : \mathbb{R}^{2n} \longrightarrow \mathbb{R}$ *be a smooth function. Show that:*

$$
X_H = \sum_{i=1}^{n}\left(\frac{\partial H}{\partial y_i}\frac{\partial}{\partial x_i} - \frac{\partial H}{\partial x_i}\frac{\partial}{\partial y_i}\right).
$$

Exercise 3.2
Show that $X_H = J\nabla H$, *where* $\nabla H = \dfrac{\partial H}{\partial x_i}\dfrac{\partial}{\partial x_i} + \dfrac{\partial H}{\partial y_i}\dfrac{\partial}{\partial y_i}.$

Let $\varphi_t : \mathbb{R}^{2n} \longrightarrow \mathbb{R}^{2n}$ be the local flow of X_H:

$$
\frac{d\varphi_t}{dt}(x) = X_H(\varphi_t(x)), \qquad \varphi_0(x) = x.
$$

If $\varphi_t(x) = \left(x_1(t),\cdots,x_1(t),y_1(t),\cdots,y_n(t)\right)$ then the equations above read:

$$
\begin{cases}
\dfrac{dx_i}{dt} = \dfrac{\partial H}{\partial y_i}\\[2mm]
\dfrac{dy_i}{dt} = -\dfrac{\partial H}{\partial x_i}.
\end{cases}
$$

These are precisely **Hamilton equations** of the motion in classical mechanics.
This system of differential equations is called a **Hamiltonian system**.

Consider for instance $H : \mathbb{R}^{2n+2} \longrightarrow \mathbb{R}$

$$H(x_1, \cdots, x_{n+1}, y_1, \cdots, y_{n+1}) = \frac{1}{2}\left(x_1^2 + \cdots + x_{n+1}^2 + y_1^2 + \cdots + y_{n+1}^2 \right).$$

Then

$$X_H = \sum_{k=1}^{n+1} \left(y_k \frac{\partial}{\partial x_k} - x_k \frac{\partial}{\partial y_k} \right).$$

Hamilton equations are:

$$\begin{cases} \dfrac{dx_k}{dt} = \dfrac{\partial H}{\partial y_k} \\[2ex] \dfrac{dy_k}{dt} = -\dfrac{\partial H}{\partial x_k}. \end{cases}$$

Putting $z_k = x_k + iy_k$, we get

$$\frac{dz_k}{dt} = \frac{dx_k}{dt} + i\frac{dy_k}{dt} = y_k - ix_k = -i(x_k + iy_k) = -iz_k$$

which have

$$z_k(t) = z_k(0)e^{-it}$$

as solutions.

The flow is generated by rotations:

$$t \longmapsto z_k(0)e^{-it}$$

on $\mathbb{R}^{2n+1}$.

This flow maps the level surface

$$S^{2n+1} = H^{-1}\left(\frac{1}{2}\right)$$

into itself.

The symplectic form ω on $\mathbb{R}^{2n}$ is preserved by the flow, so is its restrictions $\underline{\omega}$ to S^{2n+1}.

Since the rank of a skew symmetric 2-form is even, $\underline{\omega}$ is degenerate.

In fact, $\ker \underline{\omega}$ is one dimensional.

Indeed, let $X \in \ker \underline{\omega}$, $X \in T_x S^{2n+1}$ and $\omega(X, X') = 0$, $\forall \, X' \in T_x S^{2n+1}$.

But $\omega(X, X') = \langle JX, X' \rangle = 0 \ \forall \ X' \in T_x S^{2n+1}$ means $JX \in \left(T_x S^{2n+1} \right)^{\perp}$ which is one dimensional.

Now let $X' \in T_x S^{2n+1}$ and $c : (-\varepsilon, \varepsilon) \longrightarrow S^{2n+1}$ be a curve $c(0) = x$ and $X' = \frac{d}{dt} c(t)|_{t=0}$.

$$\left(i_{X_F} \omega \right)(X') = df(X') = \frac{d}{dt} f(c_t)|_{t=0} = 0$$

since $f = \text{constant}$ on S^{2n+1}.

The 2-form $\underline{\omega}$ passes to the quotient $S^{2n+1}/(\text{trajectories of } X_H) = S^{2n+1}/S^1$ to a 2-form Ω which is no longer degenerate. Hence we obtain in this way a symplectic form Ω on the complex projection spaces $\mathbb{C}P^n = S^{2n+1}/S^1$. The natural projection

$$p : S^{2n+1} \longrightarrow \mathbb{C}P^n$$

is called the **Hopf fibration**.

3.2   A characterisation of symplectic diffeomorphisms

Theorem 3.1

A diffeomorphism $\varphi : M \longrightarrow M$ *of a symplectic manifold* (M, ω) *is a* **symplectic diffeomorphism** *if and only if for all* C^∞ *function* F *on* M

$$X_{(F \circ \varphi)} = (\varphi^{-1})_* X_F.$$

The proof is an immediate consequence of the following:

Lemma 3.1

For any vector field X *and a p-form* θ *on a smooth manifold* M *and* $\varphi : M \longrightarrow M$ *a diffeomorphism, then*

$$i_{\varphi_* X} \theta = (\varphi^{-1})^* i_X (\varphi^* \theta).$$

Proof Exercise. □

Proof of the Theorem 3.1:
By Lemma 3.1

$$i_{(\varphi^{-1})_* X_F} \omega = \varphi^* i_{X_F} \left((\varphi^{-1}) * \omega \right)$$
$$= \varphi^* i_{X_F} \omega \qquad (3.2.1)$$

if and only if φ is symplectic.

But $\varphi^* i_{X_F} \omega = \varphi^* dF = d(F \circ \varphi) = i_{X_{F \circ \varphi}} \omega$.

Since ω is non-degenerate: $(\varphi^{-1})_* X_F = X_{F \circ \varphi}$ □

Corollary 3.1

Denote by Φ_f the flow of X_f, i.e $\Phi_f = (\varphi_t)$ where

$$\frac{d\varphi_t}{dt}(x) = X_f \varphi_t(x).$$

Then φ is symplectic if and only if for all smooth function $\phi_{F \circ \varphi} = \varphi^{-1} \circ \phi_F \circ \varphi$.

3.3 The Poisson bracket

Given two smooth functions $f, g : M \longrightarrow \mathbb{R}$ on a symplectic manifold (M, ω), we define a new function $\{f, g\}$ by:

$$\{f, g\} = \omega(X_f, X_g) \qquad (3.3.1)$$

called the **Poisson bracket** of f and g. Let $\mathcal{C}^\infty(M)$ denotes the space of all smooth functions on M. We just defined a map

$$\{\cdot, \cdot\} : \mathcal{C}^\infty(M) \times \mathcal{C}^\infty(M) \longrightarrow \mathcal{C}^\infty(M).$$

This operation satisfies:

1. $\{f, g\} = -\{g, f\}$,

2. It is bilinear over $\mathbb{R}$,

3. $\{f, \{g, h\}\} + \{g, \{h, f\}\} + \{h, \{f, g\}\} = 0$,

4. $\{f, u \cdot v\} = \{f, u\}v + u\{f, v\}$.

Properties (1) through (3) say that $(\mathcal{C}^\infty(M), \{\cdot,\cdot\})$ is a Lie algebra.

Property (4) is called the **Leibniz identity** and means that $\forall\, f \in \mathcal{C}^\infty(M)$

$$
\begin{aligned}
D_f: \quad \mathcal{C}^\infty(M) &\longrightarrow \mathcal{C}^\infty(M)\\
u &\longmapsto \{f,u\}
\end{aligned}
\tag{3.3.2}
$$

is a derivation.

Exercise 3.3

Prove the properties 1. – 4. above.

Let $ham(M,\omega)$ be the set of all Hamiltonian vectors field X_f, $f \in \mathcal{C}^\infty(M)$. This is a subset of $\mathfrak{X}_M$, the space of all vectors field on M. Clearly, $ham(M,\omega) \subset \mathfrak{X}_M$ is a vector subspace. Let us show it is stable for the Lie bracket of vectors fields. We know that if $X_1, X_2 \in \mathfrak{X}_M$ and θ is a p-form then

$$
i_{[X_1,X_2]}\theta = i_{X_1}L_{X_2}\theta - L_{X_2}i_{X_1}\theta.
$$

Here if $X_1 = X_f$ and $X_2 = X_g$ we have

$$
\begin{aligned}
i_{[X_f,X_g]}\omega &= i_{X_f}(L_{X_g}\omega) - L_{X_g}(i_{X_f}\omega)\\
&= -di_{X_g}(df)\\
&= -d\omega(X_g, X_f)\\
&= d(\{f,g\}).
\end{aligned}
\tag{3.3.3}
$$

Hence $[X_f, X_g]$ is a Hamiltonian vector with Hamiltonian $\{f,g\}$. Therefore $ham(M,\omega)$ is a Lie subalgebra of $\mathfrak{X}_M$.

The map

$$
\begin{aligned}
\mathcal{C}^\infty(M) &\longrightarrow ham(M,\omega)\\
f &\longmapsto X_f
\end{aligned}
\tag{3.3.4}
$$

is a Lie algebra homomorphism.

By definition it is onto and clearly its kernel is $\mathbb{R}$ if M is connected.

Let (M,ω) be a symplectic manifold and $ham(M,\omega)$ (resp. $(\mathcal{C}^\infty(M), \{\cdot,\cdot\})$) the Lie algebra consisting of Hamiltonian vectors fields (resp. of Hamiltonian functions).

Denote by ∇_s the operation $f \longmapsto X_f$. We have the following:

Proposition 3.1

If M is connected then the following short sequence

$$0 \longrightarrow \mathbb{R} \longrightarrow \left(\mathcal{C}^\infty(M), \{\cdot, \cdot\}\right) \xrightarrow{\nabla_s} ham(M, \omega) \longrightarrow 0 \qquad (3.3.5)$$

is an exact sequence.

Let us mention the following fact which is obvious in one direction:

Theorem 3.2 *(Dumortier)*

The sequence defined in equation (3.3.5) admits a section if and only if M is compact.

A vectors field X on a symplectic manifold (M, ω) is called a **symplectic vectors field** if

$$L_X \omega = 0. \qquad (3.3.6)$$

This condition means that its (local) flow φ_t preserves ω, i.e $\varphi_t^* \omega = \omega$. Condition (3.3.6) can also be written

$$d\, i_X \omega = 0$$

since $d\omega = 0$.

Let $\mathcal{L}_\omega(M)$ denote the set of all symplectic vectors fields. We know that $ham(M, \omega) \subset \mathcal{L}_\omega(M)$. If $X_1, X_2 \in \mathcal{L}_\omega(M)$, then

$$i_{[X_1, X_2]}\omega = i_{X_1} L_{X_2} \omega - L_{X_2} i_{X_1} \omega = -d i_{X_2} i_{X_1} \omega = -d\omega(X_1, X_2). \quad (3.3.7)$$

This implies that $[X_1, X_2] \in ham(M, \Omega) \subset \mathcal{L}_\omega(M)$.

Therefore $\mathcal{L}_\omega(M)$ is a sub-Lie algebra of $\mathfrak{X}_M$. The map which sends $X \in \mathcal{L}_\omega(M)$ to the cohomology class of $[i_X \omega] \in H^1(M, \mathbb{R})$ of $i_X \omega$ is a homomorphism of vectors spaces

$$c : \mathcal{L}_\omega(M) \longrightarrow H^1(M, \mathbb{R}). \qquad (3.3.8)$$

The equation (3.3.7) says that c is a Lie algebra homomorphism when $H^1(M, \mathbb{R})$ is considered as a Lie algebra with trivial bracket (an abelian Lie algebra), since c maps $[\mathcal{L}_\omega(M), \mathcal{L}_\omega(M)]$ to $0 \in H^1(M, \mathbb{R})$.

This homomorphism was first considered by Palais and Calabi [Cal70] who proved the following:

Theorem 3.3

    We have the following exact sequences of Lie algebras:

$$
\begin{array}{c}
0 \\
\downarrow \\
\mathbb{R} \\
\left(\mathcal{C}^\infty(M), [\cdot, \cdot] \right) \\
\downarrow \\
\end{array}
$$

$$
0 \;\longrightarrow\; ham(M, \mathbb{R}) \;\xrightarrow{\;i\;}\; \mathcal{L}_\omega(M) \;\xrightarrow{\;c\;}\; H^1_{DR}(M, \mathbb{R}) \;\longrightarrow\; 0
$$

$$
\begin{array}{c}
\downarrow \\
0
\end{array}
$$

(3.3.9)

where i is the natural inclusion.

Recent research consider the following questions:

Let $f_n, g_n \in \mathcal{C}^\infty_0(M)$ be sequences uniformly convergent to smooth functions f and g. Is $\{f_n, g_n\}$ converging to $\{f, g\}$?

Observe that $\{\cdot, \cdot\}$ depends on partial derivatives of f and g, which has nothing to do with the uniform convergence. So the answer may be no.

Let $f_n, g_n : \mathbb{R}^2 \longrightarrow \mathbb{R}$

$$
f_n(x, y) = \frac{u(y) \cdot \cos(nx)}{\sqrt{n}}
$$

$$
g_n(x, y) = \frac{u(y) \cdot \sin(nx)}{\sqrt{n}}
$$

(3.3.10)

where $u : \mathbb{R} \longrightarrow \mathbb{R}$ is a function with compact support. It is easy to see that f_n and g_n converge uniformly to 0 but $\{f_n, g_n\} = u(y)u'(y) \neq 0$. Under certain conditions however

$$
\{f_n, g_n\} \longrightarrow \{f, g\}
$$

where f and g are the uniform limits of f_n and g_n. For instance Cardin and Viterbo have proved:

Theorem 3.4 *[Car-Vit08]*

    Let f_n, g_n be sequences of smooth functions converging uniformly to smooth functions f and g. Then

$$
\{f_n, g_n\} = 0 \quad \forall \, n \Longrightarrow \{f, g\} = 0.
$$

This result has been generalized by Humilière, Cardin and Viterbo, etc. [Hum08], [Car-Vit08].

These kind of results are parts of what is called the "**symplectic rigidity**": the interplay between the C^0-topology (uniform convergence) and symplectic properties (which are C^∞ objects).

The question of the rigidity of the Poisson bracket has recently gained a renewed interest [Car-Vit08], [Ent-Pol10], [Ent-Pol-Zap07], [Pol-Dan14].

The symplectic rigidity is one of the objects of the Epilogue.

3.4   Integrable Hamiltonian systems

Let H be a C^∞ function on a symplectic manifold (M, ω). The system of the first order differential equations

$$\dot{x} = X_H$$

is a Hamiltonian system. We denote this system by (M, ω, H)

Proposition 3.2

Let F and H be two C^∞ functions on a symplectic manifold (M, ω). Then

$$\{F, H\} = 0 \iff F \text{ is constant along the flow of } X_H.$$

    Proof

Let φ_t be the flow of X_H. Then

$$
\begin{aligned}
\frac{d}{dt}(F \circ \varphi_t) &= \varphi_t^* \left(L_{X_H} F \right) \\
&= \left(\varphi_t^* dF \right)(X_H) \\
&= \left(\varphi_t^* \left(i_{X_F} \right) \omega \right)(X_H) \\
&= (\varphi^* \omega)\left(X_F, X_H \right) \\
&= \{F, H\} \circ \varphi_t.
\end{aligned}
\tag{3.4.1}
$$

The assertion of the proposition follow.   □

A function F such that $\{F, H\} = 0$ is called **first integral** or **the constant** of the motion (φ_t).

Two functions F and G are said to be in **involution** if $\{F, G\} = 0$.

Definition 3.1

Let (M, ω) be a $2n$-dimensional symplectic manifold. A hamiltonian system (M, ω, H) is said to be **completely integrable** *if there exists a system of n functions $\{f_1, f_2, \cdots f_n\}$ with $f_1 = H$, which are pairwise in involution, namely $\{f_i, f_j\} = 0$ for all i, j and which are linearly independent on a dense open subset U of M, i.e on U, one has*

$$df_1 \wedge df_2 \wedge \cdots \wedge df_n \neq 0 \quad (*).$$

We have the following famous fact:

Theorem 3.5 *(Arnold-Liouville) [Lib-Mar87]*

Let (M, ω, H) be a completely integrable hamiltonian system with $\{H = f_1, f_2, \cdots f_n\}$ first integral in involution which are independent on a dense subset $U \subseteq M$ $(*)$

Let $F = (f_1, f_2, \cdots f_n) : M \longrightarrow \mathbb{R}^n$ and let $a \in \mathbb{R}^n$ be regular point of F. We have

1. *If $F_a = F^{-1}(a)$ is compact and connected then F_a is a torus T^n.*

2. *In that case, there is a diffeomorphism φ of neighbourhood $\mathcal{U}_a$ of F_a onto $T^n \times D^n$ where D^n is an open disk in $\mathbb{R}^n$ centred at the origin such that*

   (a) $\left(\varphi_{|F_a}\right)(F_a) = T^n \times \{0\}$;

   (b) $\varphi^*(\omega_{T^n}) = \omega_{|\mathcal{U}_a}$.

   Here ω^{T^n} is the restriction to $T^n \times D^n$ of the natural symplectic form $d(\lambda_{(T^n)})$ of the cotangent bundle $T^*T^n = T^n\mathbb{R}^n$.

   (c) Let pr_1, pr_2 be the projections on first and second factor in $T^n \times D^n$

   $$pr_1 : T^n \times D^n \longrightarrow T^n \qquad and \qquad pr_2 : T^n \times D^n \longrightarrow D^n$$

   and denote by

   $$\theta_a : pr_1 \circ \varphi = (\theta_1, \cdots, \theta_n) \qquad \text{(the "angle coordinates")}$$
   $$q_a : pr_2 \circ \varphi = (q_1, \cdots, q_n) \qquad \text{(the "action coordinates").}$$

*Then the hamiltonian system (M, ω, H) restricted to $\mathcal{U}_a$ is "quasi-periodic",
i.e the flow φ_t of X_H is given by*

$$\varphi_t(\theta_1, \cdots, \theta_n, q_1, \cdots, q_n) = (\theta_1 + t\sigma_1, \cdots, \theta_n + t\sigma_n)$$

for some fixed $(\sigma_1, \cdots, \sigma_n)$ $\left(\varphi_t \text{ is independent of the action coordinates}\right)$.

Remark 3.1

*The compact and connected submanifold T_a is Lagrangian, since
$\left(\varphi_{|F_a}\right)(F_a) = T^n \times \{0\}$ and $d(\lambda_{T^n})$ vanishes there.*

*Hence $F : \mathcal{U}_a \longrightarrow U_a$ is a Lagrangian fibration (called a Duisteremaat
fibration), where U_a is a neighborhood of $a \in \mathbb{R}^n$.*

In Appendix B, we will give a full treatment of the integrability in contact geometry, following an unpublished paper by A. Banyaga and P. Molino. An outline of this paper appeared in [Ban99].

3.5 Hamiltonian diffeomorphisms

We saw that if $\varphi_t : M \longrightarrow M$ is an isotopy (a smooth family of diffeomorphisms with $\varphi_0 = id$) then we get a smooth family of vector fields $\dot{\varphi}_t$ defined by

$$\dot{\varphi}_t(x) = \frac{d\varphi_t}{dt}\left(\varphi_t^{-1}(x)\right). \tag{3.5.1}$$

Let (M, ω) be a symplectic manifold. An isotopy φ_t is called a **symplectic isotopy** if $\dot{\varphi}_t$ is a symplectic vector field, i.e

$$L_{\dot{\varphi}_t}\omega = di_{\dot{\varphi}_t}\omega = 0.$$

The isotopy φ_t is called a **Hamiltonian isotopy** if there exists a smooth family of functions $F = F(x, t)$ such that

$$\dot{\varphi}_t = X_F$$

where X_F is the family of vector field defined by

$$i_{X_F}\omega = dF_t.$$

We defined:
$Symp(M, \omega)_0$ to be the set of all diffeomorphisms φ with compact support

such that there exists a symplectic isotopy with compact support γ_t with $\gamma_1 = \varphi$ and $Ham(M, \omega)$ the set of all diffeomorphisms φ with compact support such that there exists a Hamiltonian isotopy γ_t with $\gamma_1 = \varphi$.

Exercise 3.4
   Show that the kernel ker S *of the Calabi homomorphism is equal to* $Ham(M, \omega)$.

Remark 3.2
   Every $\varphi \in \mathrm{Symp}(M, \omega)$ *supported in a ball belongs to* $Ham(M, \omega)$. *It follows from the proof of Boothby theorem that if* (M, ω) *is arcwise connected,* $Ham(M, \omega)$ *acts p-transitively on* M.

Exercise 3.5
   Show that if (M, ω) *is a compact symplectic manifold, any* $f \in$ $Ham(M, \omega)$ *sufficiently close to the identity in* C^1*-topology has a fixed point.*

   This follows easily from Moser theory. This simple remark was the starting point of the "Arnold conjecture" of the sixties [Arn65].

Arnold conjecture
   Let (M, ω) be a compact symplectic manifold, the number of fixed points of a Hamiltonian diffeomorphism φ whose graph intersects transversally the diagonal is bounded from below by the sum of the Betti numbers of M.

$$\#(\Gamma_\varphi \cap \Gamma_{id}) \geqslant \sum_{k=0}^{n} \dim H^k_{DR}(M).$$

This conjecture has been a driving force in research in symplectic topology. It is nowadays considered to be "almost" completely settled.
   We close this section by reformulating the fundamental result in Theorem 2.5:

Theorem 3.6 *(Banyaga [Ban78])*
   Let (M, ω) *be a compact symplectic manifold. Then* $Ham(M, \omega)$ *is a simple group.*

   This theorem says that there is no non-trivial homomorphism from $Ham(M, \omega)$ onto an abelian group.

3.6 Poisson manifolds

We used the symplectic form ω of a symplectic manifold to construct the Poisson bracket $\{\cdot,\cdot\}$ and a Lie algebra structure on $\mathcal{C}^\infty(M)$. We are going now to shift our interest on this Lie algebra structure.

Definition 3.2

A **Poisson structure** π on a smooth manifold M is the data of a Lie algebra structure $\{\cdot,\cdot\}$ on $\mathcal{C}^\infty M$ such that $\forall\, f$,

$$D_f : \mathcal{C}^\infty M \longrightarrow \mathcal{C}^\infty M, \quad D_f(u) = \{f,u\}$$

is the derivation of $\mathcal{C}^\infty M$.

The couple (M,π) is called a **Poisson manifold**.

By definition a symplectic form on a symplectic manifold defines a Poisson structure. But there are much more Poisson manifolds which are not symplectic.

Example 3.1

Let (N,ω) be a symplectic manifold. Then for any smooth manifold P, $M = N \times P$ is a **Poisson** manifold.

Indeed for $f,g \in \mathcal{C}^\infty(M)$ define

$$\{f,g\}(x,y) = \{f_y,g_y\}(x)$$

where $f_y(x) = f(x,y), \quad g_y(x) = g(x,y)$.

This example can be generalized: If M is a manifold equipped with a foliation where each leaf is a symplectic manifold, we define a bracket like above leaf-by-leaf. This Poisson structure is called a **Dirac bracket**.

Example 3.2

Let $\mathcal{G}$ be a finite dimensional Lie algebra and $M = \mathcal{G}^*$. If $f \in \mathcal{C}^\infty(M)$ and $\theta \in \mathcal{G}^*$,

$$d_\theta f : T_\theta \mathcal{G}^* \simeq \mathcal{G}^* \longrightarrow T_{f(\theta)}\mathbb{R} = \mathbb{R}$$

is then an element of $(\mathcal{G}^*)^* = \mathcal{G}$. If $[\cdot,\cdot]$ is a bracket on $\mathcal{G}$, then we define $\{f,g\}(\theta) = \theta\Big(\big[d_\theta f, d_\theta g\big]\Big).$

Exercise 3.6

Verify that $\{\cdot, \cdot\}$ is a Poisson structure on M.

This structure is called a **Kostant-Kirillov-Souriau (KKS) struc-ture**. It plays an important role in the theory of group representations [Kir76].

Exercise 3.7

Exhibit a formula for the KKS structure on

$$\mathcal{G} = \left\{ \begin{pmatrix} 0 & x & y \\ -x & 0 & z \\ -y & -x & 0 \end{pmatrix} \mid x, y, z \in \mathbb{R} \right\}.$$

Chapter 4

Group actions

This chapter is a brief introduction to symplectic/hamiltonian actions. For more details, the reader may consult the original papers, or the books by Guillemin and Sternberg [Gui-Ste77, Gui-Ste90].

4.1 Basic definitions

Definition 4.1

A Lie group is a manifold G which is equipped with a group structure such that the group operations

$$\begin{array}{ccc} G \times G & \longrightarrow & G \\ (a,b) & \longrightarrow & a \cdot b \end{array} \qquad and \qquad \begin{array}{ccc} G & \longrightarrow & G \\ g & \longrightarrow & g^{-1} \end{array}$$

are smooth maps.

Fixing $g \in G$ we get smooth maps

$$L_g : \begin{array}{ccc} G & \longrightarrow & G \\ a & \longrightarrow & g \cdot a \end{array} \qquad and \qquad R_g : \begin{array}{ccc} G & \longrightarrow & G \\ a & \longrightarrow & a \cdot g \end{array}$$

called the left and right translation. We denote by $\mathcal{G}$ the set of all L_g-invariant vector fields on G and call it the **Lie algebra** of G. The Lie algebra $\mathcal{G}$ is isomorphic with $T_e G$, the tangent space to G at the identity element e of G.

Definition 4.2

There is a map $Exp : \mathcal{G} \longrightarrow G$ which is an isomorphism of $\mathcal{G}$ with a neighborhood of the identity e in G. This map is called the exponential map.

Definition 4.3

*A **smooth (left) action** of a Lie algebra G on a smooth manifold M is a smooth map $A : G \times M \longrightarrow M$ such that*

$$\begin{array}{ccc} A(g_1 \cdot g_2, x) & = & A\big(g_1, A(g_2, x)\big) \\ A(e, x) & = & x \end{array}$$

for all g_1, g_2 in G and $x \in M$.

Usually one denotes $A(g,x)$ by $g \cdot x$. In this case the equations above read

$$(g_1 g_2) \cdot x = g_1 \cdot (g_2 \cdot x), \quad e \cdot x = x.$$

A (left) action A on M gives rise to a homomorphism ϕ_A from G to the group $\mathrm{Diff}^\infty(M)$ of diffeomorphisms of M:

$$g \longmapsto \phi_A(g): \quad \begin{matrix} M & \longrightarrow & M \\ x & \longmapsto & g \cdot x. \end{matrix}$$

The action also gives a Lie algebra homomorphism

$$\varphi_A: \quad \mathcal{G} \longrightarrow \mathrm{Vect}(M)$$
$$\xi \longmapsto \varphi_A(\xi) = \frac{d}{dt}_{|t=0}\Big(\phi_A\big(Exp(t\xi)\big)\Big)(x).$$

We simply denote $\varphi_A(\xi)$ by $\underline{\xi}$ and call it the **fundamental vector field** of $\xi \in \mathcal{G}$.

Let (M,ω) be a symplectic manifold. An action A of a Lie group $\mathcal{G}$ is said to be a **symplectic action** if

$$\big(\phi_A(g)\big)^*\omega = \omega$$

for all $g \in G$.

The action is said to be **hamiltonian** if $\phi_A(g)$ is a hamiltonian diffeomorphism.

A hamiltonian action of a Lie group G is usually defined by giving a Lie algebra homomorphism

$$\mu: \quad \mathcal{G} \longrightarrow C^\infty(M)$$
$$\xi \longmapsto \underline{\mu}(\xi) = F_A$$

from the Lie algebra $\mathcal{G}$ to the Poisson algebra $\big(C^\infty(M), \{\cdot,\cdot\}\big)$ of M where

$$i_{\underline{\xi}}\omega = dF_A.$$

Definition 4.4

The **momentum map** *of hamiltonian action* $\underline{\mu} : \mathcal{G} \longrightarrow C^{\infty}(M)$ *is the function* $\mu : M \longrightarrow \mathcal{G}^*$ *where* $\mathcal{G}^*$ *is the dual of* $\mathcal{G}$ *defined by:*

$$\mu(x)(\xi) = F_A(x).$$

Definition 4.5

A Lie group acts on itself: for each $g \in G$, *let*

$$\begin{array}{rcl} A(g) : & G & \longrightarrow & G \\ & a & \longmapsto & g \cdot a \cdot g^{-1}. \end{array}$$

The tangent map at e *of the map above:* $T_e\big(A(g)\big) : T_e G = \mathcal{G} \longrightarrow T_e G = \mathcal{G}$ *is called the* **adjoint action** *of* G *on* $\mathcal{G}$ *and it is denoted* $Ad_g : \mathcal{G} \longrightarrow \mathcal{G}$.

4.2 Examples

4.2.1 Examples of Lie group

1. $\mathbb{R}^n$ (with the addition);

2. $S^1, T^n = \underbrace{S^1 \times S^1 \times \cdots \times S^1}_{n \text{ times}}$;

3. $GL(n, \mathbb{R})$, the group of $n \times n$-invertible matrices;

4. $\mathcal{O}(n) = \{A \in Gl(n, \mathbb{R}) , A^t A = I\}$ the set of orthogonal $n \times n$ matrices;

5. $SO(n) = \{A \in O(n) , \det A = 1\}$.

4.2.2 Examples of group actions

Let $G = \mathbb{R}$. A group action of $\mathbb{R}$ on a smooth manifold M is called a "1-**parameter group**". It is defined as the flow of a complete vector fields X:

$$\mathbb{R} \ni t \longmapsto \varphi_t : M \longrightarrow M$$

such that

$$\left\{ \begin{array}{rcl} \frac{d}{dt}\varphi_t(x) & = & X_{\varphi_t(x)} \\ \varphi_0(x) & = & x \end{array} \right. .$$

The action is symplectic if $L_X\omega = 0$.

If $i_X\omega$ is exact then the action is hamiltonian.

If φ_t is 2π-periodic, i.e $\varphi_1 = \varphi_{2\pi}$ we say that we have an S^1-action.

1. For instance, consider the action of S^1 on $\mathcal{C}^n \equiv \mathbb{R}^{2n}$

$$\big(u, (z_1, z_2 \cdots, z_n)\big) \longmapsto (uz_1, uz_2 \cdots, uz_n).$$

   The action is hamiltonian and is generated by the vector field

$$\sum \left(-y_j\frac{\partial}{\partial x_j} + x_j\frac{\partial}{\partial y_j}\right).$$

   The momentum map is

$$\mu(z_1, z_2, \cdots, z_n) = \frac{1}{z}\sum_{j=1}^{n}|z_j|^2.$$

2. Similarly $T^n = \{t_1, t_2, \cdots, t_n \in \mathbb{C}, \ |t_i| = 1\}$ acts on $\mathbb{C}^n = \mathbb{R}^{2n}$ by

$$(t_1, t_2, \cdots, t_n) \cdot (z_1, z_2, \cdots, z_n) = (t_1 z_1, t_2 z_2, \cdots, t_n z_n).$$

   The momentum map is

$$\mu(z_1, \cdots, z_n) = \left(\frac{1}{2}\big(|z_1|^2 + |z_2|^2 + \cdots + |z_n|^2\big) + \text{constant}\right) \in \mathbb{R}^n \simeq (\mathbf{t^n})^*$$

   where $(\mathbf{t^n})^*$ is the Lie algebra of T^n.

3. Let $H_1, H_2, \cdots, H_k : M \longrightarrow \mathbb{R}$ be k-functions with compact support and which commute, i.e

$$\{H_i, H_j\} = 0, \quad \forall i, j.$$

   The flows $\Phi = (\varphi_t^1, \varphi_t^2, \cdots, \varphi_t^k)$ define a hamiltonian action of the torus T^k on M whose momentum map is

$$\mu = (H_1, H_2, \cdots, H_k) : M \longrightarrow \mathbb{R}^k \simeq (\mathbf{t^k})^*.$$

Exercise 4.1

Suppose that u and v in $\mathcal{C}^\infty(M,\mathbb{R})$ are invariant under the action of G on the symplectic manifold (M,ω), i.e

$$u(g \cdot x) = u(x) \quad and \quad v(g \cdot x) = v(x) \qquad \forall g \in G.$$

Show that $\{u,v\}$ is also G-invariant.

We note the following important result:

Theorem 4.1 *(Noether Theorem)*

Let (M,ω) be a symplectic manifold with a hamiltonian action of a Lie group G and let $\mu : M \longrightarrow \mathcal{G}^*$ be the momentum map. Let $f : M \longrightarrow \mathbb{R}$ be a smooth function which is G invariant. Then the momentum map μ is constant on the trajectories of X_f.

Proof

Let $f : M \longrightarrow \mathbb{R}$ be a function invariant by the action of G and let $\underline{X}$ be the fundamental vector field of $X \in \mathcal{G}$. Then $L_{\underline{X}}f = 0$.

Now the momentum map $\mu : M \longrightarrow \mathcal{G}^*$ is

$$\langle\, \mu(x), X \,\rangle = F_{\underline{X}}(x)$$

where $i_{\underline{X}}\omega = dF_{\underline{X}}$.

We need to prove that $F_{\underline{X}}$ is invariant under the flow of f:

$$
\begin{aligned}
L_{X_f} F_X &= i_{X_f} dF_X \\
&= i_{X_f} i_{\underline{X}}\omega = -i_{\underline{X}} i_{X_f}\omega \\
&= -i_{\underline{X}} df \\
&= -L_{\underline{X}} f = 0.
\end{aligned}
\qquad (4.2.1)
$$

4.3 Symplectic reduction

We start with a few definitions. If we have a (left) action of a Lie group G on a smooth manifold M, we call the **orbit of** G through a point $x \in M$ the set

$$\{g \cdot x \ \ g \in G\} = G \cdot x.$$

The **stabilizer** of a point $p \in M$ is the subgroup

$$G_p = \{g \in G \;\; g \cdot p = p\}.$$

The action is said to be **transitive** if there is just one orbit; it is said to be **free** if all stabilizers are trivial $\{e\}$.

Let $\sim$ be the orbit equivalence relation:

$$p \sim q \longleftrightarrow p \text{ and } q \text{ are on the same orbit.}$$

Denote by $M/\sim = M/G$ the orbit space. We have a natural projection

$$\begin{array}{rccc} \pi : & M & \longrightarrow & M/G \\ & p & \longmapsto & \text{orbit through } p. \end{array} \qquad (4.3.1)$$

Definition 4.6

 *A **principal G-bundle** is a manifold P with a smooth map $\pi : P \longrightarrow M$ such that*

 1. G acts (on the left) freely on P.

 2. $M \simeq P/\sim$ is the orbit space and π is the natural projection.

 3. There is an open cover $\mathcal{U} = (U_i)$ of M such that for each U_i there is a smooth map $\varphi_i : \pi^{-1}(U_i) \longrightarrow U_i \times G$ satisfying

 (a) $\varphi_i(p) = \big(\pi(p), s_{U_i}(p)\big)$

 (b) and $s_{U_i}(g \cdot p) = g \cdot p = g s_{U_i}(p)$ for all $p \in \pi^{-1}(U_i)$.

The map $\pi : M \longrightarrow M/G$ above is a **principal G-bundle**. We are now in position to state the reduction theorem.

Theorem 4.2 *(Marsden-Weinstein-Mayer Reduction theorem)*

 Let (M,ω) be a symplectic manifold equipped with a hamiltonian left action of a Lie group G with momentum map $\mu : M \longrightarrow \mathcal{G}^*$.

 Suppose G acts freely on $\mu^{-1}(0) \subset M$. Then

 1. $M_{red} := \mu^{-1}(0)/G$ is a smooth manifold.

 2. The natural projection $\pi : \mu^{-1}(0) \longrightarrow M_{red}$ is a principal G-bundle.

3. *There is a symplectic form ω_{red} on M_{red} such that*

$$i^*\omega = \pi^*\omega_{red}$$

where i is the inclusion $i : \mu^{-1}(0) \subset M$. We have the following diagram:

$$
\begin{array}{ccc}
\mu^{-1}(0) & \overset{i}{\hookrightarrow} & (M,\omega) \\
\pi_{\mu^{-1}(0)} \downarrow & & \downarrow \pi \\
\mu^{-1}(0)/G & \overset{j}{\hookrightarrow} & (M,\omega)/G.
\end{array}
$$

Definition 4.7
 The pair (M_{red}, ω_{red}) is called the **reduction of** (M,ω) with repect to G, μ.

Example 4.1
 The function $H : \mathbb{R}^{2n+2} \longrightarrow \mathbb{R}$

$$H(x_1, x_2, \cdots, x_{n+1}, y_1, y_2, \cdots, y_{n+1}) = \frac{1}{2}(x_1^2 + \cdots + x_{n+1}^2 + y_1^2 + \cdots y_{n+1}^2)$$

is the momentum map of an S^1-action on $\mathbb{R}^{2n+2}$ and $\mu^{-1}(0)/S^1 = \mathbb{C}P^n$.

4.4 Convexity theorem

We finish this introduction to group actions by a beautiful theorem due to Atiyah [Ati82] and independently to Guillemin-Sternberg [Gui-Ste77, Gui-Ste90].

Theorem 4.3 *(Atiyah, Guillemin-Sternberg)*
 Let (M,ω) be a compact symplectic manifold of dimension $2n$ equipped with a (left) hamiltonian action of the torus T^n with momentum map $\mu : M \longrightarrow \mathbb{R}^n$. Then

1. *the level sets of μ are connected*

2. *the image of μ is a convex subset of $\mathbb{R}^n$*

3. *the image of μ is the convex hull of images of fixed points of the action.*

It has been proved by Delzant [Del88] that the polytop $\mu(M) \subset \mathbb{R}^n$ determines completely (M, ω), i.e from the knowledge of $\mu(M)$ we may recon-
struct (M, ω) and the T^n-action.

In Appendix B, on the complete integrability in contact geometry, we
give complete proof of some generalization of these ideas to contact geometry.

Chapter 5

Contact manifolds

Contact geometry is the odd dimensional counterpart of symplectic geometry.

In his book, "Contact geometry and Wave propagation" [Arn89] Arnold wrote: *Every theorem in symplectic geometry may be formulated as a contact geometry theorem and an assertion in contact geometry may be translated in the language of symplectic geometry.*

Definition 5.1

 *A **contact form** on a $2n + 1$-dimensional manifold M is a 1-form α such that*

$$\alpha \wedge (d\alpha)^n$$

is a volume form.

Recall that a symplectic form on a $2n$-dimensional manifold M is a 2-form ω such that ω^n is a volume form.

A **contact manifold** is a couple (M, α) of a differentiable manifold M and a contact form α on M.

A contact manifold is oriented by the volume form $\alpha \wedge (d\alpha)^n$.

We start this section with a long list of contact forms. The reader may skip some examples in Section 5.1.

5.1 Examples

5.1.1 Basic examples

1. Let $(x_1, \cdots, x_n, y_1, \cdots, y_n, z)$ be the coordinates on the Euclidean space $\mathbb{R}^{2n+1}$, the following 1-forms are contact forms:

$$\alpha_1 = \sum_{i=1}^{n} x_i dy_i \pm dz$$

$$\alpha_2 = \frac{1}{2}\sum_{i=1}^{n}(x_i dy_i - y_i dx_i) \pm dz$$

$$\alpha_3 = \sum_{i=1}^{n} \rho^2 d\theta_i \pm dz \text{ where } \rho^2 = x_i^2 + y_i^2 \text{ and } \tan\theta_i = \frac{y_i}{x_i}.$$

The form α_1 is the local prototype of any contact form. Namely we have, like in symplectic geometry, the contact Darboux theorem:

Theorem 5.1 *(Darboux theorem)*

Let α be a contact form on a $(2n+1)$-dimensional manifold M. For each point $x_0 \in M$, there exists an open neighborhood U of x_0 and a chart $\varphi : U \longrightarrow \mathbb{R}^{2n+1}$ with $\varphi(x_0) = 0$ and

$$\varphi^*(\alpha_1) = \alpha_{|U}$$

where α_1 is the standard contact form

$$\sum_{i=1}^{n} x_i dy_i + dz$$

of $\mathbb{R}^{2n+1}$.

2. Some contact forms on $\mathbb{R}^3$.

   The following forms are contact on $\mathbb{R}^3$:

   $$\beta_1 = (\cos x)dz + (\sin x)dy;$$
   $$\beta_2 = (\cos y)dx + (\sin y)dz;$$
   $$\beta_3 = (\cos z)dx + (\sin z)dy.$$

   These forms are invariant under translation and hence descend to the torus T^3.

Exercise 5.1

Show that in cylindrical coordinates ρ, θ, z in $\mathbb{R}^3$ the following 1-forms are contact forms.

$$\gamma_1 = \rho(\sin\rho)d\theta + (\cos\rho)dz$$

$$\gamma_2 = (1 - \rho^4)dz + z\rho^2 d\theta$$
$$\gamma_3 = (\cos \rho^2 + \sin \rho^2)d\theta.$$

These forms also descend to T^3.

Exercise 5.2

Show that for all n

$$
\begin{aligned}
\alpha_n \;=\; & \cos\left[\left(\frac{\pi}{4} + n\pi(x_2^2 + y_2^2)\right)\right][x_1 dy_1 - y_1 dx_1] \\
& + \sin\left[\left(\frac{\pi}{4} + n\pi(x_2^2 + y_2^2)\right)\right][x_2 dy_2 - y_2 dx_2] \quad (5.1.1)
\end{aligned}
$$

are contact forms on S^3 (Lutz).

5.1.2 More examples

1. We have already seen some contact forms on T^3. In fact, we have the following theorem due to Martinet [Mar70].

   Theorem 5.2

   Every oriented 3-dimensional manifold admits a contact form.

2. On T^5:
   the 1-form

   $$\alpha = \varphi_1 d\theta_4 + \varphi_2 d\theta_3 + \sin\theta_2 \cos\theta_2 d\theta_1 - \sin\theta_1 \cos\theta_1 d\theta_2 + \cos\theta_1 \cos\theta_2 d\theta_3$$

   where $\varphi_1 = \sin\theta_1 \cos\theta_3 - \sin\theta_2 \sin\theta_3$ and $\varphi_2 = \sin\theta_1 \sin\theta_3 - \sin\theta_2 \cos\theta_3$ is a contact form on T^5 *[Lutz]*.

3. T^{2n+1} has a contact form [Bou74].

4. S^{2n+1}: On $\mathbb{R}^{2n+2}$ we consider the form

   $$\theta = \frac{1}{2}\left(\sum_{i=1}^{n+1} x_i dy_i - y_i dx_i\right) \quad (5.1.2)$$

   and $i : S^{2n+1} \longrightarrow \mathbb{R}^{2n+2}$ the inclusion. One easily show that $\alpha = i^*\theta$ is a contact form on S^{2n+1}.

5. If (M_1, α_1) and (M_2, α_2) are two contact manifolds, then $Q = M_1 \times M_2 \times \mathbb{R}$ has the following contact form: let $\pi_i : Q \longrightarrow M_i$, $t : Q \longrightarrow \mathbb{R}$ be the projections; for any non-zero numbers a and b

$$\alpha = at\pi_1^* \alpha_1 + b\pi_2^* \alpha_2 \qquad (5.1.3)$$

is a contact form on Q.

6. The cosphere bundle

Let $T^*M \smallsetminus \{0\}$ be the cotangent space minus the zero section. The group $\mathbb{R}^+$ of strictly positive numbers acts on $T^*M \smallsetminus \{0\}$ by:

$$k \in \mathbb{R}^+, \underline{a} = (x, \alpha_x) \in T^*M \smallsetminus \{0\}; \quad k \cdot \underline{a} = (x, k\alpha_x).$$

This action is free and proper; therefore

$$\pi : T^*M \smallsetminus \{0\} \longrightarrow P = T^*M \smallsetminus \{0\} / \mathbb{R}^+$$

is a smooth fiber bundle. Let $\sigma : P \longrightarrow T^*M \smallsetminus \{0\}$ be a global section of π and let λ_M be the Liouville 1-form of T^*M, then $\alpha = \sigma^* \lambda_M$ is a contact form on P. Using a riemannian metric, we can identify P with the cosphere bundle

$$S_M^* = \left\{ (x, \alpha_x), \|\alpha_x\| = 1 \right\}$$

and hence get a contact form on S_M^*. For instance, we get a contact form on each of these manifolds:

$$
\begin{array}{ccc}
\mathbb{R}P^3 & \text{as} & S_{S^2}^* \\
T^3 & \text{as} & S_{T^2}^* \\
T^n \times S^{n-1} & \text{as} & S_{T^n}^* \\
PSL(2, \mathbb{R}) & \text{as} & S_{H^2}^*.
\end{array}
$$

Exercise 5.3

Find a diffeomorphism of $\mathbb{R}^3$ pulling back $\alpha = \cos(z)dx + \sin(z)dy$ to $\beta = dz - ydx$.

Exercise 5.4

Show that

$$\rho: \quad \mathbb{R}^{2n+1} \quad \longrightarrow \quad \mathbb{R}^{2n+1}$$
$$(x, y, z) \quad \longmapsto \quad (x+y, y-x, z + \tfrac{1}{2}|x|^2 - \tfrac{1}{2}|y|^2)$$

pulls back $\alpha_1 = \sum_{i=1}^{n} x_i dy_i + dz$ *to* $\beta_1 = \sum_{i=1}^{n} (x_i dy_i - y_i dx_i) + dz.$

Exercise 5.5

Show that the composition rule $\perp$ *on* $\mathbb{R}^{2n+1}$:

$$(x, y, z) \perp (u, v, w) = \big(x+u, y+v, z+w + (x \cdot v - y \cdot u)\big) \qquad (5.1.4)$$

turns $\mathbb{R}^{2n+1}$ *into a group (the* **Heisenberg group***).*

Show that for all $a \in \mathbb{R}^{2n+1}$, *the translation*

$$R_a : (x, y, z) \mapsto (x, y, z) \perp a$$

preserves β_1, *i.e* $R_a^* \beta_1 = \beta_1$, *where* $\beta_1 = \sum_{i=1}^{n} (x_i dy_i - y_i dx_i) + dz.$

Exercise 5.6

Show that the map $(\rho, \theta, z) \longmapsto (\rho, \theta - \rho^2 z, \frac{z}{2}(1 + \rho^4))$ *pulls* γ_2 *to* $\alpha = x dy - y dx + dz.$

5.2 Relation with symplectic manifolds

5.2.1 Contactization of symplectic manifold

If (M, ω) is a symplectic manifold with an exact symplectic form, $\omega = d\lambda$, then

$$P = M \times \mathbb{R}$$

is a contact manifold with contact form $\alpha = \pi^* \lambda + dt$ where

$$t : M \times \mathbb{R} \longrightarrow \mathbb{R} \qquad \text{and} \qquad \pi : M \times \mathbb{R} \longrightarrow M$$

are canonical projections.

1. The basic examples α_1, α_2 and α_3 are of this nature.

2. If $M = T^*X$ is the cotangent bundle of a manifold X with symplectic form $\omega_X = d\lambda_X$ where λ_X is the Liouville form, the $\alpha = \pi^* \lambda_X + dt$ is a contact form on $M \times \mathbb{R}$.

5.2.2   Symplectization SP of a contact manifold (P, α)

The space SP is the 1-dimensional subbundle of T^*P whose fibers over x are non-zero multiples of $\alpha(x)$.

Hence SP is a trivial $\mathbb{R}^+$-bundle over P.

Using this $\mathbb{R}^+$-actions, we can view SP as $P \times \mathbb{R}^+$.

As a submanifold of T^*P, SP inherits the symplectic form $d\lambda_P$, i.e $\omega = i^*(d\lambda_P)$ where $i : SP \longrightarrow T^*P$ is the natural inclusion. Within the trivialization $SP \simeq P \times \mathbb{R}^+$, the symplectic form ω becomes

$$\omega = d(e^t \alpha) = e^t(d\alpha - \alpha \wedge dt).$$

Let $j : P \longrightarrow SP \subset T^*P$ be the inclusion of the graph of α. We have:

$$j^* \lambda_P = \alpha.$$

Therefore

$$dj^* \lambda_P = j^* d\lambda_P = j^* \omega_P$$

or

$$j^* \omega = d\alpha.$$

5.2.3   Hypersurfaces of contact type in a symplectic manifold

Definition 5.2 *(Weinstein)*

A hypersurface $N^{2n-1} \overset{j}{\hookrightarrow} (M^{2n}, \omega)$ embedded in a symplectic manifold is a **hypersurface of contact type** *if there exists a contact form α on N such that $j^*\omega = d\alpha$.*

Example 5.1

1. *Any contact manifold (P, α) is a hypersurface of contact type in its symplectization.*

2. *The sphere S^{2n+1} with its standard form (example 4) is the hypersurface of contact type in $\mathbb{R}^{2n+2}$.*

3. *The cosphere bundle S_M^* is an example of hypersurface of contact type in $T^*M \smallsetminus \{0\}$.*

4. *The natural inclusion*

$$T^3 \xhookrightarrow{j} T^4$$

is not a hypersurface of contact type in $\left(T^4, \omega = dx_1 \wedge dx_2 + dx_3 \wedge dx_4\right)$
since $j^*\omega = dx_1 \wedge dx_2$ *which is not an exact form.*

To recognize hypersurfaces of contact type, we can use the following characterization:

Theorem 5.3 *(Weinstein)*

A hypersurface M in a symplectic manifold (P, Ω) is a hypersurface of contact type if and only if there exists a vector field Z defined on a neighborhood U of M in P which is transverse to M and such that $L_Z \Omega = \Omega$ on U.

5.3 The Reeb field of a contact form

Since $\alpha \wedge (d\alpha)^n \neq 0$ everywhere, $d\alpha$ has rank $2n$; its kernel is 1-dimensional. Let $\xi_x \in \ker d\alpha$ and choose $v_1, \cdots, v_{2n} \in T_x M$ to complete a basis $(\xi_x, v_1, \cdots, v_{2n})$ of $T_x M$ so that

$$\alpha_x \wedge (d\alpha_x)^n \big(\xi_x, v_1, \cdots, v_{2n}\big) = 1.$$

One has

$$
\begin{aligned}
1 &= \alpha_x \wedge (d\alpha_x)^n \big(\xi_x, v_1, \cdots, v_{2n}\big) \\
&= \alpha_x(\xi_x)(d\alpha_x)^n \big(v_1, \cdots, v_{2n}\big) \\
&\quad + \sum_{i=1}^{2n} (-1)^i \alpha_x(v_i)(d\alpha_x)^n \big(v_1, \cdots, v_{i-1}, \xi_x, v_{i+1}, \cdots, v_{2n}\big) \\
&= \alpha_x(\xi_x)(d\alpha_x)^n \big(v_1, \cdots, v_{2n}\big)
\end{aligned}
$$

since $\xi_x \in \ker d\alpha_x$ each term in the summation is zero.

Thus $\alpha_x(\xi_x) \neq 0$ for all x. Hence we can normalize ξ_x and define a vector field, we denote again by ξ such that

$$
\begin{aligned}
\alpha(\xi) &= 1 \\
i_\xi d\alpha &= 0.
\end{aligned}
\tag{5.3.1}
$$

This vector field is unique and is called the **Reeb field** of α.

Example 5.2

1. *The Reeb field of $\alpha_1 = \sum\limits_{i=1}^{2n} x_i dy_i + dz$ is $\xi = \dfrac{\partial}{\partial z}$.*

2. *The Reeb field of $\alpha = (\cos x_3)dx_1 + (\sin x_3)dx_2$ is*

$$\xi = (\cos x_3)\frac{\partial}{\partial x_1} + (\sin x_3)\frac{\partial}{\partial x_2}.$$

3. *The Reeb field of the contact form α on S^{2n+1} in Example 4 of Section 5.1.2, is:*

$$\xi = \sum_{i=1}^{n} x_i \frac{\partial}{\partial y_i} - y_i \frac{\partial}{\partial x_i}.$$

We saw in the discussion after Exercise 3.2 that all of its orbits are periodic with period one.

Definition 5.3

   A contact form α on a compact manifold M is said to be a **regular contact form** *if all the orbits of its Reeb field are periodic with period one.*

Example 5.3

1. *The forms $\alpha = i^*\theta$ on S^{2n+1} is a regular contact form.*

2. *The contact form $\alpha = (\cos z)dx + (\sin z)dy$ on T^3 is not regular.*

   Indeed the orbit through $(0,0,\frac{\pi}{3})$ of the Reeb field
   $\xi = (\cos z)\dfrac{\partial}{\partial x} + (\sin z)\dfrac{\partial}{\partial y}$ *is*

$$t \longmapsto \left(x = \frac{1}{2}t, y = \frac{\sqrt{3}}{2}t, z = \frac{\pi}{2}\right)$$

   which is an irrational flow on the T^2-torus $z = \frac{\pi}{3}$.

In fact, Blair has proved [Bla02]:

Theorem 5.4

   No torus admits a regular contact form.

5.3.1   Contact dynamics

A contact manifold (M, α) comes equipped with an important dynamical system

$$\dot{x} = \xi$$

where ξ is the Reeb field of α.

Observe that its flow preserves the contact form since:

$$L_\xi \alpha = d i_\xi \alpha + i_\xi d\alpha = 0 + 0 = 0.$$

5.3.2   The Weinstein's conjecture

In 1979, Weinstein made the following conjecture [Wei79]:

"The Reeb field of form α on a compact contact manifold M (with $H^1(M, \mathbb{R}) = 0$) must have at least one periodic orbit."

Like the Arnold conjecture (on the existence of fixed points of Hamiltonian diffeomorphism) this conjecture has been a driving force in contact and symplectic topology.

There is an abundant literature on this conjecture.

The first important result was proved by Viterbo [Vit87] for contact manifold which are hypersurfaces of contact type in $(\mathbb{R}^{2n}, \omega_0)$.

The idea was to transform the problem into a variational problem.

The most recent result was a proof of Weinstein conjecture in some 3-dimensional manifolds by Hofer-Zehnder using Gromov pseudoholomorphic curves.

In between, there have been some partial interesting result: for instance by Banyaga [Ban90] proved the conjecture for forms which are C^0-close to regular contact forms and Rukimbira [Ruk95] for contact forms the Reeeb field of which is a Killing vector field with respect to some contact metric structure.

Before, Weinstein conjecture was known to be true if the contact manifold is a convex hypersurface of contact type (Weintein 1978) or starshapped (Rabinowicz 1978).

Exercise 5.7

Find a perturbation of the contact form on S^{2n+1} above so that the resulting Reeb field has only 2 periodic orbits.

5.3.3   Regular contact flows

We now prove the following results:

Theorem 5.5 *(Boothby-Wang) [Boo-Wan78]*

Let (M, α) be a compact regular contact manifold and B the orbit space of the Reeb flow. Then B is a symplectic manifold whose symplectic form Ω has integral periods. Moreover, the projection

$$\pi : M \longrightarrow B$$

is a principal S^1-bundle and we have $\pi^*\Omega = d\alpha$.

Conversely, if (B, Ω) is a symplectic manifold where Ω has integral periods, there exists a principal S^1-bundle

$$\pi : M \longrightarrow B$$

over B where M has a regular contact form α such that $\pi^*\Omega = d\alpha$.

Proof

Let α be a regular contact form on a compact manifold M. The Reeb field of α defines then a free circle action on M. Let $B = M/S$ be the orbit space. This is a smooth manifold and the natural projection $\pi : M \longrightarrow B$ is a principal S^1-bundle. The 2-form $d\alpha$ is a basic form, i.e it is invariant under the contact flow, namely $L_\xi(d\alpha) = dL_\xi\alpha = 0$. Therefore there is a 2-form Ω on B such that
$$\pi^*\Omega = d\alpha.$$

The 2-form Ω is non-degenerate: for all $b \in B$, the tangent space $T_bB \simeq$ Orthogonal complement V_p of ξ on T_pM where $p \in \pi^{-1}(b)$, $(d\alpha)^n \neq 0$ on V_p.

Obviously $d\Omega = 0$, so Ω is a symplectic form on B.

Let $\mathcal{U} = (U_i)$ be an open cover of M, where each U_i is the domain of a distinguished coordinates chart: $\varphi_i : U_i \longrightarrow \mathbb{R}^{2n+1}$ ($\dim M = 2n + 1$), i.e if $\varphi_i(x) = (z, x_1(x), \cdots, x_{2n}(x))$, then an orbit of Z through a point in U_i has coordinates $(z, a_1, \cdots, a_{2n})$ where $a_1, \cdots, a_{2n}$ are constant numbers.

Let $V_i = \pi(U_i)$. Then $\{V_i\}$ is an open cover of B and we may introduce the following trivializations

$$
\begin{aligned}
\psi_i : \quad S^1 \times V_i \quad &\longrightarrow \quad \pi^{-1}(V_i) \\
(t, p) \quad &\longmapsto \quad t \cdot \big(s_i(p)\big),
\end{aligned}
$$

where $s_i : V_i \longrightarrow \pi^{-1}(V_i)$ is a local section.

On $V_i \cap V_j$, we have:

$$
s_j = \gamma_{ij} \cdot s_i,
$$

where $\gamma_{ij} : V_i \cap V_j \longrightarrow S^1$ are the transition functions of the bundle.

Let $w_i = s_i^* \alpha$ on V_i we have:

$$
dw_i = ds_i^* \alpha = s_i^* (\pi^* \Omega) = (\pi \circ s_i)^* \Omega = \Omega,
$$

since $\pi \circ s_i = id$. On the other hand

$$
w_j = s_j^* \alpha = w_i + d\widetilde{\gamma}_{ij},
$$

where

$$
\widetilde{\gamma}_{ij} : V_i \cap V_j \longrightarrow \mathbb{R}
$$

is a lift of $\gamma_{ij} : V_i \cap V_j \longrightarrow S^1$; which exists since $V_i \cap V_j$ is contractible. Because $\gamma_{ij} = \gamma_{ik}\gamma_{kj}$, one has

$$
\widetilde{\gamma}_{ij} + \widetilde{\gamma}_{jk} + \widetilde{\gamma}{ki} \in \mathbb{Z}.
$$

But the Ceĥ representation of the cohomology class of Ω is just given by the cocycle

$$
\bar{c}_{ijk} = \widetilde{\gamma}_{ij} + \widetilde{\gamma}_{jk} + \widetilde{\gamma}_{ki}
$$

which is an integer. We conclude that the symplectic form Ω on B has integral periods [Bot-Tu82].

Conversely, let (B, Ω) be a symplectic manifold, where the symplectic form Ω has integral periods, i.e, its cohomology class $[\Omega]$ belong to $H^2(M, \mathbb{Z})$. Choose a good open cover $\{V_i\}$ of B (The V_i and any finite intersection are contractible open sets.) By Poincaré's lemma, there exists 1-form ω_i on V_i such that

$$
\Omega_{|V_i} = d\omega_i
$$

and on $V_i \cap V_j$ (which is contractible).

There are functions f_{ij} such that

$$\omega_i - \omega_j = df_{ij}$$

since

$$d\omega_i - d\omega_j = \Omega_{|V_i} - \Omega_{|V_j} = 0 \qquad \text{on} \qquad V_i \cap V_j.$$

On $V_i \cap V_j \cap V_k$,

$$d(f_{ij} + f_{jk} + f_{ki}) = (\omega_i - \omega_j) + (\omega_j - \omega_k) + (\omega_k - \omega_i) = 0.$$

The constant $c_{ijk} = f_{ij} + f_{jk} + f_{ki}$ is a Cech cocycle representing the cohomology class $[\Omega]$ of Ω. It is an integer by assumption, hence

$$\exp(2\pi c_{ijk}) = 1 \in S^1.$$

Now, defining

$$\bar{f}_{ij} : \quad V_i \cap V_j \quad \longrightarrow \quad S^1$$
$$p \quad \longmapsto \quad \exp\left(2\pi f_{ij}(p)\right)$$

we see that

$$\bar{f}_{ij}\bar{f}_{jk}\bar{f}_{ki} = e$$

or equivalently $\bar{f}_{ik} = \bar{f}_{ij}\bar{f}_{jk}$.

Therefore $\bar{f}_{ij}$ is a 1-cocycle with values in S^1. This cocycle determines, as it is well known, a principal S^1-bundle

$$\pi : M \longrightarrow B.$$

Let us recall the construction.

M is a disjoint union $\cup V_i \times S^1$ where $(x, g) \in V_i \times S^1$ and $(x', g') \in V_j \times S^1$ are identified if and only if $x = x' \in V_i \cap V_j$ and $g' = \gamma_{ij}(x)g$.

The map $\pi : M \longrightarrow B$ is the projection on the first factor. On M, the 1-forms ω_i fit together to produce a 1-form α on M, which is a regular contact form and the orbits of its Reeb field are the orbits of the circle action on M. Moreover, it is clear that $\pi^*\Omega = d\omega$. $\qquad\square$

Corollary 5.1

Let M be a compact 3-dimensional manifold with a free-action of S^1 which defines a non-trivial fibration; then M carries a regular contact form.

Proof

Let B be the quotient space M/S^1. Since the action is free, $\pi : M \longrightarrow B$ is a principal S^1-bundle over the 2-dimensional compact manifold B.

Let Ω be a volume form on B such that

$$\int_B \Omega = 1.$$

The cohomology class $a \in H^2(B, \mathbb{Z})$ of Ω determines an S^1-bundle

$$\pi' : M' \longrightarrow B$$

where M' has a regular contact form α' with $d\alpha' = \pi'^*\Omega$.

Since π and π' have the same Chern class a, they are isomorphic [Bot-Tu82], i.e there exists a diffeomorphism $h : M \longrightarrow M'$ with $\pi' \circ h = \pi$. Therefore $\alpha = h^*\alpha'$ is a regular form on M. $\square$

The S^1-principal bundles we just constructed are called the **prequantization bundles** of the symplectic manifold where the symplectic form has integral periods.

Example 5.4

The Hopf fibration $\pi : S^{2n+1} \longrightarrow \mathbb{C}P^n$ is the prequantization bundle of $(\mathbb{C}P^n, \Omega)$.

Contact manifold associated with a non-integral symplectic form

This procedure allow to construct contact manifolds starting with compact symplectic manifold with not necessarily integral symplectic forms. Let Ω be any symplectic form on a compact manifold B. By the theorem of Hodge – de-Rham, $\Omega = \Omega_0 + du$ where Ω_0 is a harmonic form. Let $h = \{h_1, \cdots, h_p\}$ be a basis of harmonic 2-forms. Then

$$\Omega_0 = \sum_{i=1}^{p} r_i h_i$$

and the numbers r_i represent the periods of Ω. Approximate r_i by rational numbers $\frac{n_i}{m_i}$ and consider

$$\Omega' = \sum_{i=1}^{p} \frac{n_i}{m_i} h_i + du.$$

If the approximations $\dfrac{n_i}{m_i}$ are good enough, Ω' is still a symplectic form with rational periods. Multiplying Ω' by $\lambda = m_1 m_2 \cdots m_p$,

$$\omega'' = \lambda \Omega'$$

is a symplectic form with integral periods.

Now using the later, we may construct a principal S^1-bundle $\pi : M \longrightarrow B$, where M carries a regular contact form.

5.4 Contact structures

Definition 5.4

A contact structure on a manifold M of dimension $2n + 1$ is a subbundle $E \subset TM$ of dimension $2n$ (a hyperplane) such that each $x \in M$ has neighborhood U and a contact form α_U on U such that $E_{|_U}$ is the kernel of α_U.

The Reeb field ξ_U of α_U is thus transverse to E.

If in the definition above, the contact form α_U is a global contact form, then the contact structure E is said to be "**co-oriented**".

For each $x \in M$ consider a basis $\{X_1, \cdots, X_{2n}\}$ of E_x and complete it by ξ_x to get a basis of $T_x M$, such that

$$\alpha_U \wedge (d\alpha_U)^n (\xi_x, X_1, \cdots, X_{2n}) = 1.$$

Since $\alpha_U(X_i) = 0$, we have:

$$\alpha_U \wedge (d\alpha_U)^n (\xi_x, X_1, \cdots, X_{2n}) = \alpha_U(\xi_x)(d\alpha_U)^n (X_1, \cdots, X_{2n}).$$

Hence $(d\alpha_U)^n (X_1, \cdots, X_{2n}) = 1$.

The restriction of $d\alpha_U$ to E is non-degenerate.

Definition 5.5

*Two contact forms α and α' such that $\alpha' = \lambda \alpha$ where λ is a nowhere zero functions are said to be **equivalent**.*

Clearly two equivalent contact forms define the same contact structure E

$$E = \ker(\lambda \alpha) = \ker \alpha.$$

If a diffeomorphism $\varphi : M \longrightarrow M$ pulls back a contact form α to $\lambda\alpha$, i.e $\varphi^*\alpha = \lambda\alpha$ where λ is a nowhere zero function, then φ preserves the contact distribution of α. Such a diffeomorphism is called a **contact diffeomorphism** or a **contactomorphism**.

Example 5.5

Let $\alpha = dz - ydx$ on $\mathbb{R}^3$. Since

$$h : \quad \mathbb{R}^3 \quad \longrightarrow \quad \mathbb{R}^3$$
$$(x, y, z) \quad \longmapsto \quad (z\cos x - y\sin x, -z\sin x - y\cos x, -x)$$

pulls back β to α, these two forms define the same contact structure on $\mathbb{R}^3$.

Example 5.6

A deep theorem of Bennequin [Ben83] asserts that on $\mathbb{R}^3$, the contact form

$$\rho \sin \rho (d\theta) + (\cos \rho) dz$$

and the standard contact form

$$\alpha dz - ydx$$

define different contact structures.

Definition 5.6

*Let α be a contact form on M, a 1-form θ on M which vanished along the Reeb field ξ is called **semi-basic**.*

It can be viewed as a section of E^*, the dual bundle of E.

Since $d\alpha$ is non-degenerate on E, such a 1-form θ determines uniquely a vector field H_θ which is a section of E such that

$$i_{H_\theta} d\alpha = \theta.$$

For instance if $f : M \longrightarrow \mathbb{R}$ is a smooth function. Then

$$\theta = \left(i_\xi df\right)\alpha - df$$

is semi-basic. Hence it defines a unique section H_f of E^*, such that

$$i_{H_f} d\alpha = (i_\xi df)\alpha - df.$$

Exercise 5.8

Let α be a contact form on M with Reeb field ξ. For any nowhere zero function λ on M show that the Reeb field ξ_λ of $\alpha_\lambda = \lambda\alpha$ is given by

$$\xi_\lambda = \mu\xi + H_\mu$$

where $\mu = \dfrac{1}{\lambda}$.

Remark 5.1

This shows that the Reeb fields of equivalent contact forms may be very different.

5.5   Two basic theorems

Like in symplectic geometry we have a Darboux theorem asserting that $(\mathbb{R}^{2n+1}, \alpha_1)$,

$$\alpha_1 = \sum_{i=1}^{2n} x_i dy_i + dz$$

is the local model of any contact form on M (Theorem 5.1).

Proof of Darboux theorem

Let $x \in M$ and V be a neighborhood of 0 in $T_x M$, of the form $V = V_0 \times \,]-\varepsilon, \varepsilon[$ where V_0 is a neighborhood $0 \in E_x$.

The geodesics coordinates gives a diffeomorphisms φ from V to a neighborhood U of $x \in M$.

For each $t \in \,]-\varepsilon, \varepsilon[$ the restriction of $d\alpha$ to $U'_t = \varphi(V_0 \times \{t\})$ is a closed 2-form of maximum rank, i.e a symplectic form.

By the (symplectic) Darboux theorem, each point $u \in U'_t$ has a neighborhood $\tilde{U}_t$ of 0 in U'_t and coordinates

$$\Big(x_1(u), x_2(u), \cdots, x_n(u), y_1(u), y_2(u), \cdots, y_n(u), z\Big)$$

such that

$$d\alpha_{|_{\tilde{U}_t}} = \sum_{i=1}^{n} dx_i \wedge dy_i.$$

Remark that the form $d\alpha$ is invariant in the z-direction.

Hence on $U = U' \cap \left(\bigcup_{t \in]-\varepsilon, \varepsilon[} \tilde{U}_t \right)$ we have:

$$d(\alpha - \sum_{i=1}^{n} x_i dy_i) = 0.$$

Therefore

$$\alpha = \sum x_i dy_i + dw$$

for some function w on U.

Since $\alpha \wedge (d\alpha)^n \neq 0$, the function $\left(x_1(u), \cdots, x_n(u), y_1(u), \cdots, y_n(u), w \right)$ are functional independent and hence make the desired coordinates. $\square$

The next important fact is a thoerem proved by Gray [Gra59] and reproved by Martinet [Mar70] using Moser path method. This reminds Moser's theorem on equivalent of symplectic forms.

Theorem 5.6 *(Gray-Martinet)*

Let α_t be a smooth path of contact forms on a compact manifold M. There exists a smooth family φ_t of diffeomorphisms of M such that $\varphi_0 = id$ and a family of functions u_t such that

$$\varphi_t^* \alpha_t = u_t \alpha. \tag{5.5.1}$$

Proof

The 1-form

$$\beta_t = \dot{\alpha}_t(\xi_t)\alpha_t - \dot{\alpha}_t$$

is semi-basic.

There exists a unique X_t section of E such that

$$i_{X_t} d\alpha_t = \beta_t.$$

Since $i_{X_t}\alpha_t = 0$, we see that

$$L_{X_t}\alpha = \beta_t.$$

Let φ_t be the family of diffeomorphisms defined by X_t, i.e

$$\begin{cases} \frac{d}{dt}\varphi_t(x) &= X_t(\varphi_t(x)) \\ \varphi_0 &= id \end{cases}.$$

Then:

$$\frac{d}{dt}(\varphi_t^* \alpha_t) = \varphi^*(L_{X_t}\alpha_t + \dot{\alpha}_t) = \varphi^*(\beta_t + \dot{\alpha}_t)$$
$$= \varphi_t^*(\dot{\alpha}_t(\xi)\alpha_t). \tag{5.5.2}$$

We now determine u_t such that $\varphi_t^* \alpha_t = u_t \alpha_0$.

If we denote by $\dot{u}_t = \dfrac{\partial u_t}{\partial t}$ we have

$$\frac{d}{dt}(\varphi_t^* \alpha_t) = \dot{u}_t \alpha_0$$
$$= \dot{u}_t \frac{\varphi_t^* \alpha_t}{u_t}. \tag{5.5.3}$$

Hence by (5.5.2), we have

$$\varphi_t^*(\dot{\alpha}_t(\xi))\varphi_t^* \alpha_t = \frac{\dot{u}_t}{u_t}\varphi_t^* \alpha_t$$

or

$$\dot{\alpha}_t(\xi) \circ \varphi_t = \frac{d}{dt}(\ln u_t)$$

which can be integrated to give

$$u_t = \exp\left(\int_0^t (\dot{\alpha}_t(\xi) \circ \varphi_s)ds\right).$$

The isotopy φ_t and the family of functions u_t satisfy the equation (5.5.1).
$\square$

Stability of a contact structures

If α is a contact form on a smooth manifold M and β is 1-form which is C^1-close to zero then

$$\alpha' = \alpha + \beta$$

is still a contact form since the condition $\alpha' \wedge (d\alpha')^n \neq 0$ is an open condition.

Hence $\alpha_t = \alpha + t\beta$ is a smooth family of contact forms. By Gray-Martinet Theorem, there exists an isotopy φ_t and a family of smooth functions u_t such that

$$\varphi_t^* \alpha_t = u_t \alpha_0.$$

Therefore

$$\varphi_1^* \alpha' = u_0 \alpha$$

i.e the forms α and α' define equivalent contact structures.

Remark 5.2

In his thesis V. Colin (1998) proved that on a compact 3-dimensional manifold two contact forms which are only C^0-close are isotopic.

5.6 Contactomorphisms

Let (M, α) be a contact manifold with contact form α. A **contactomorphism** of (M, α) is a diffeomorphism φ of (M, α) which preserves the contact form structure $E = \ker \alpha$. Such a diffeomorphism is characterized by this property:

$$\varphi^* \alpha = \lambda \alpha$$

for some nowhere zero function λ. When $\lambda = 1$, we say that φ is a **strictly contact diffeomorphism**.

Observe that the function λ such that $\varphi^* \alpha = \lambda \alpha$ is unique, since $\mu \alpha = \varphi^* \alpha = \lambda \alpha$ implies that $(\lambda - \mu)\alpha = 0$. since $\alpha(X) \neq 0$ for all X; we see that $\lambda = \mu$.

We have the following easy fact.

Proposition 5.1

The set $\mathrm{Diff}(M, \alpha)$ of all contactomorphisms of M forms a group (under the composition of maps). The subset $\mathrm{Diff}_\alpha(M)$ of strictly contact diffeomorphisms is a subgroup of $\mathrm{Diff}(M, \alpha)$ but is not a normal subgroup.

Proof

For $\varphi \in \mathrm{Diff}(M, \alpha)$, let λ_φ denotes the function such that $\varphi^* \alpha = \lambda_\varphi \alpha$. We have:

1. $\lambda_{id} = 1$

2. $\lambda_{\varphi^{-1}} = \dfrac{1}{\lambda_\varphi \circ \varphi^{-1}}$

3. $\lambda_{\varphi \circ \psi} = \lambda_\varphi \circ \psi \cdot \lambda_\psi.$

The fact that $\text{Diff}(M,\alpha)$, and $\text{Diff}_\alpha(M)$ are groups follows.

Now we see that if $h \in \text{Diff}_\alpha(M)$, then for $\varphi \in \text{Diff}(M,\alpha)$,

$$\lambda_{(\varphi \circ h \circ \varphi^{-1})} = (\lambda_\varphi \circ h \circ \varphi^{-1}) = \frac{\lambda_\varphi(h \circ \varphi^{-1})}{\lambda_\varphi \circ \varphi^{-1}} \neq 1$$

unless $h = id$ which means that $\text{Diff}_\alpha(M)$ is not a normal subgroup of $\text{Diff}(M,\alpha)$. $\qquad\square$

Example 5.7

1. $M = \mathbb{R}^{2n+1}$, $\alpha = \sum_{i=1}^{n} y_i dx_i + dz$. *For any non-zero number* μ,

$$\varphi_\mu: \quad \begin{array}{ccc} \mathbb{R}^{2n+1} & \longrightarrow & \mathbb{R}^{2n+1} \\ (x,y,z) & \longmapsto & (\mu x, \mu y, \mu^2 z) \end{array}$$

satisfies $\varphi_\mu^* \alpha = \mu^2 \alpha$.

2. *The translations:*

$$T_a : (x,y,z) \in \mathbb{R}^{2n+1} \longmapsto T_a(x,y,z) = (x+a, y, z)$$

where $a \in \mathbb{R}^n$ *and*

$$\tau_\lambda : (x,y,z) \in \mathbb{R}^{2n+1} \longmapsto \tau_\lambda(x,y,z) = (x,y,z+\lambda)$$

where $\lambda \in \mathbb{R}$ *are strictly contact diffeomorphisms.*

Let $\mathcal{L}(M,\alpha)$ be the set of all vector fields X on M such that

$$L_X \alpha = \mu \alpha$$

for some function μ on M and the set $\mathcal{L}_\alpha(M)$ of vector fields Y such that

$$L_Y \alpha = 0.$$

The local flow of a vector field X with $L_X \alpha = 0$, integrate to a flow $\varphi_t \in \text{Diff}_\alpha(M)$.

Suppose now that φ_t is the local flow of Y with $L_Y \alpha = \mu\alpha$, then

$$\varphi_t^* \alpha = u_t \alpha \qquad (5.6.1)$$

where $u_t = \exp\left(\int_0^t (\mu \circ \varphi_s)ds\right)$. Hence $\varphi_t \in \text{Diff}(M, \alpha)$.

To prove (5.6.1), we write:

$$\dot{u}_t \alpha = \frac{d}{dt}(\varphi_t^* \alpha) = \varphi_t^*(L_X \alpha)$$
$$= (\mu \circ \varphi_t)(u_t \alpha) \qquad (5.6.2)$$

or

$$\frac{u_t'}{u_t} = \frac{d}{dt}(\ln u_t) = \mu \circ \varphi_t$$

then (5.6.1) follows.

Proposition 5.2

The spaces $\mathcal{L}(M, \alpha)$ and $\mathcal{L}_\alpha(M)$ are Lie subalgebras of the Lie algebra of all vectors fields.

Proof

We only need to show that these vector fields are closed under the Lie bracket on the space of all vector fields.

If $L_X \alpha = u_X \alpha$, $L_Y \alpha = u_Y \alpha$ then:

$$L_{[X,Y]}\alpha = L_X L_Y \alpha - L_Y L_X \alpha$$
$$= (X \cdot u_Y - Y \cdot u_X)\alpha. \qquad (5.6.3)$$

This shows that $[X, Y]$ belongs to $\mathcal{L}(M, \alpha)$.

Clearly, if X and Y are in $\mathcal{L}_\alpha(M)$ the $[X, Y] \in \mathcal{L}_\alpha(M)$. $\qquad \square$

The Lie algebra $\mathcal{L}(M, \alpha)$ is called the Lie algebra of **contact vector fields** and $\mathcal{L}_\alpha(M)$ the Lie algebra of **strictly contact vector fields**.

Let $\mathcal{C}^\infty(M)$ denote the space of smooth functions. We have the following important fact.

Theorem 5.7

Let (M, α) be a contact manifold with contact form α. The map

$$I : \quad \mathcal{L}(M, \alpha) \longrightarrow \mathcal{C}^\infty(M)$$
$$X \longmapsto i_X \alpha = \alpha(X) \qquad (5.6.4)$$

is an isomorphism of vector spaces.

Proof

Clearly the map I is linear.

We construct the inverse of I.

Given $f \in C^\infty(M)$, the 1-form $\Theta_f = (df)(\xi) - df$ is semi-basic, i.e $\Theta_f(\xi) = 0$. (Here ξ is the Reeb field of α). Therefore there is a unique 1-form on E ($= \ker \alpha$), H_f such that

$$i_{H_f} d\alpha = \Theta_f.$$

Now, set $Y_f = H_f + f\xi$. We have

$$\alpha(Y_f) = \alpha(H_f) + f\alpha(\xi) = f$$

(recall that H_f vanishes on α).

$$
\begin{aligned}
L_{Y_f}\alpha &= di_{Y_f}\alpha + i_{Y_f}d\alpha \\
&= df + \Theta_f \\
&= df + (\xi \cdot f)\alpha - df \\
&= (\xi \cdot f)\alpha.
\end{aligned}
\tag{5.6.5}
$$

We conclude that

$$L_{Y_f}\alpha = (\xi \cdot f)\alpha$$

and $i_{Y_f}\alpha = f$. Hence $Y_f \in \mathcal{L}(M, \alpha)$. $\qquad\square$

5.6.1 Applications

1. Given $f, g \in C^\infty(M)$ we can define a bracket $[f, g]$ by the following formula:
$$[f, g] := \alpha\Big([Y_f, Y_g]\Big) = I([Y_f, Y_g]).$$

   With this bracket, $C^\infty(M)$ becomes a Lie algebra.

   This bracket is called the **Jacobi bracket**.

   The pair $\Big(C^\infty(M), [\cdot, \cdot]\Big)$ is a Lie algebra which is a contact version of the Poisson structure in symplectic geometry.

Exercise 5.9

Give an explicit formula for the bracket $[f, g]$ for two functions $f, g \in C^\infty(\mathbb{R}^{2n+1})$, where $\mathbb{R}^{2n+1}$ is equipped with $\alpha = \sum\limits_{i=1}^{n} y_i dx_i + dz$.

2. Contactization of vector field

Each general vector field X gives rise to a unique contact vector field $\widetilde{X}$:

$$\tilde{X} = I^{-1}\big(\alpha(X)\big) \in \mathcal{L}(M,\alpha).$$

If $X \in \mathcal{L}(M,\alpha)$, i.e $X = I^{-1}\big(\alpha(X)\big)$, we see that $\widetilde{X} = X$.

Fragmentation property

If $\mathcal{U} = (U_j)_{j \in J}$ is an open cover of the contact manifold (M,α) and $(\lambda_j)_{j \in J}$ a partition of unity subordinate to $\mathcal{U}$. For each $X \in \mathcal{L}(M,\alpha)$;
$X = \sum\limits_{j \in J} \lambda_j X.$

$$X = \widetilde{X} = \sum_{j \in J} I^{-1}(\lambda_j X) = \sum_{j \in J} \widetilde{X}_j \tag{5.6.6}$$

where $\widetilde{X}_j \in \mathcal{L}(M,\alpha)$ and has support in U_j. (This is the "**fragmentation property**".)

□

5.6.2 Some properties of the group of contactomorphisms

Theorem 5.7 shows that the group $\mathrm{Diff}(M,\alpha)$ of contactomorphisms of a contact manifold (M,α) is very large. Indeed, any smooth function with compact support gives rise to a contact vector field. We may integrate and get a contactomorphism.

In fact, $\mathrm{Diff}(M,\alpha)$ may be viewed as an *"infinite dimensional Lie group"* with Lie algebra $\mathcal{L}(M,\alpha)$, the Lie algebra of contact vector fields.

We refer to [Ban97], for the following facts:

Theorem 5.8

The group $\mathrm{Diff}(M,\alpha)$ is locally connected by differentiable arcs.

Let $\mathrm{Diff}(M,\alpha)_c$ be its subgroup formed by elements with compact support and $\mathrm{Diff}(M,\alpha)_0$ its identity component. Then $\mathrm{Diff}(M,\alpha)_0$ is made of $\varphi \in \mathrm{Diff}(M,\alpha)_c$ which are smooth isotopic to the identity, i.e there exists a smooth path $\rho_t \in \mathrm{Diff}(M,\alpha)_c$ with $\rho_0 = id$ and $\rho_1 = \varphi$.

Exercise 5.10

Prove that $\text{Diff}(M, \alpha)_0$ has the "**fragmentation property**", i.e given $\varphi \in \text{Diff}(M, \alpha)_0$ and an open cover $\mathcal{U} = (U_i)_{i \in I}$ of M, there exists $\varphi_1, \varphi_2,$ $\cdots, \varphi_N$ where φ_i has support in U_i for all $i \in I$ and

$$\varphi = \varphi_1 \varphi_2 \cdots \varphi_N.$$

Hint Use the fragmentation property of the corresponding family of contact vector field and the fact that $\text{Diff}(M, \alpha)_0$ is locally connected by arcs,

We need the fact

Theorem 5.9

There exists a neighborhood of the identity in $\text{Diff}(M, \alpha)_0$ which is "smoothly" diffeomorphic to a neighborhood of zero in some vector space (of contact vector fields).

Recently, Rybicki proved [Ryb10]:

Theorem 5.10

The group $\text{Diff}(M, \alpha)_0$ is a simple group.

It was known by Banyanga-McInerney [Ban-Ine95], that the commutator subgroup $\left[\text{Diff}(M, \alpha)_0, \text{Diff}(M, \alpha)_0\right]$ is simple. Rybicki had just proven that $\text{Diff}(M, \alpha)_0$ is perfect.

Recall that a group G is perfect if it is equal to its commutator subgroup $[G, G]$.

Exercise 5.11

Prove the contact version of Boothby theorem in symplectic geometry:

Let (M, α) be a connected contact manifold. Then $\text{Diff}(M, \alpha)$ acts p-transitively on M, i.e given the sets $(x_1, \cdots, x_n)$, and $(y_1, \cdots, y_n)$ of distinct points, there exists $\varphi \in \text{Diff}(M, \alpha)$ such that $\varphi(x_i) = y_i$ $\forall i = 1, 2, \cdots, n$.

5.7 Contact metric structures

Here we define the analog of adapted metrics for symplectic forms.

Definition 5.7

A **contact metric structure** *on a contact manifold* (M, α) *is a couple* (g, ϕ) *where* g *is a Riemannian on* M *and* ϕ *is a* $(1,1)$*-tensor field* ϕ : $TM \longrightarrow TM$ *satisfying:*

1. $\phi(\xi) = 0$,

2. $\phi^2(X) = -X + \alpha(X)\xi$ *where* ξ *is the Reeb field of* α,

3. $d\alpha(X, Y) = g(X, \phi Y)$,

4. $g(X, Y) = g(\phi X, \phi Y) + \alpha(X)\alpha(Y)$.

Property 4. implies that $\alpha(X) = g(X, \xi)$.

In the metric g, the Reeb field ξ has norm 1 and if X is a section of $E = \ker \alpha$

$$g(X, \xi) = \alpha(X) = 0.$$

We view $\mathfrak{X}_M$ as $\mathfrak{X}_M = \mathcal{V} \oplus \mathcal{H}$ where $\mathcal{V} = \mathbb{R}\xi$ and $\mathcal{H} = $ sections of E. In fact we have the decomposition of any vector fields X as

$$X = X_v + X_h \qquad \text{where} \qquad X_v = (i_X \alpha)\xi \qquad X_h = X - X_v.$$

Theorem 5.11

Any contact manifold (M, α) *admits infinitely many contact metric structures and all of them are homotopic.*

Proof

Let $E = \ker \alpha$ be the contact distribution. Since $(E, d\alpha_{|_E})$ is a symplectic vector bundle, the construction in Section 1.5 provides a complex structure J_0 on E compatible with $d\alpha$, i.e

$$d\alpha(J_0 X, J_0 Y) = d\alpha_{|_E}(X, Y)$$

for all sections X, Y of E and $(X, Y) \longmapsto d\alpha_{|_E}(X, J_0 Y)$ is a Riemannian metric g_0 on E. We know that we recover $d\alpha_{|_E}$ by:

$$d\alpha_{|_E}(X, Y) = g_0(J_0 X, Y).$$

We extend the complex structure J_0 on E to a $(1,1)$-tensor $\phi : TM \longrightarrow TM$ by

$$\begin{cases} \phi(\xi) &= 0 \\ \phi(X) &= J_0 X \qquad \text{for all} \quad X \in E \end{cases} \tag{5.7.1}$$

and g_0 to a Riemannian metric on M as follows:

$$g(X,Y) \quad = \quad g_0\big(\phi(X),\phi(Y)\big) + \alpha(X)\alpha(Y). \tag{5.7.2}$$

It is clear that g is a Riemannian metric.

Properties *1*, *3* and *4* are obvious from definitions. It is also easy to check the property *2*: $\phi^2(X) = \phi(\phi(X)) = \phi(J_0 X_h) = J_0(J_0 X_h)$ since $J_0 X_h$ is horizontal; hence

$$\phi^2(X) = J^2 X_h = -X_h = -\Big(X - (i_X\alpha)\xi\Big).$$

Therefore we just constructed a contact metric structure using a compatible complex structure J_0 for $d\alpha|_E$. Since these compatible structures are infinite and are all homotopic, the theorem follows.   □

Example 5.8

A contact metric structure for $\alpha = dz - ydx$ on $\mathbb{R}^3$.

The Reeb field of α is $\xi = \frac{\partial}{\partial z}$.
The contact distribution E is spanned by

$$V_1 = \frac{\partial}{\partial y}$$

and

$$V_2 = y\frac{\partial}{\partial z} + \frac{\partial}{\partial x}.$$

The adapted basis of $T_x\mathbb{R}^3$ is then $\Big\{V_1, V_2, V_3\Big\}$ where $V_3 = \xi = \frac{\partial}{\partial z}$.

The adapted metric satisfies:

$$g(V_i, V_j) = \delta_{ij}.$$

Let $e_1 = \frac{\partial}{\partial x}$, $e_2 = \frac{\partial}{\partial x}$ and $e_3 = \frac{\partial}{\partial x}$ be the natural basis.

$$\begin{aligned}
e_1 &= V_2 - y\frac{\partial}{\partial x} = V_2 - yV_3 \\
e_2 &= V_1 \\
e_3 &= V_3.
\end{aligned}$$

An immediate calculation gives

$$
\begin{aligned}
g(e_1, e_1) &= 1 + y^2, & g(e_1, e_2) &= 0 \\
g(e_2, e_3) &= 1 + y^2, & g(e_1, e_3) &= -y \\
g(e_2, e_2) &= g(e_3, e_3) = 1.
\end{aligned}
$$

$$(5.7.3)$$

Hence the matrix of g is:

$$
\begin{pmatrix}
1 + y^2 & 0 & -y \\
0 & 1 & 0 \\
-y & 0 & 1
\end{pmatrix}.
$$

$$(5.7.4)$$

More generally, with the coordinates $(x_1, \cdots, x_n, y_1, \cdots, y_n, z)$, the contact form

$$
\alpha = dz - \sum_i y_i dx_i
$$

with the Reeb field $\xi = \dfrac{\partial}{\partial z}$.

The contact distribution E is spanned by $\dfrac{\partial}{\partial y_1}, \dfrac{\partial}{\partial y_n}, \left[y_i \dfrac{\partial}{\partial z} + \dfrac{\partial}{\partial x_i} \right]$, and a contact metric g has the following matrix (in standard basis)

$$
G = \begin{pmatrix}
\delta_{ij} + y_i y_j & 0 & -y_i \\
0 & \delta_{ij} & 0 \\
-y_i & 0 & 1
\end{pmatrix}.
$$

Chapter 6

Solutions of selected exercises

Solution 6.1 *(to Exercise 1.1)*

Let us show that A defines a symplectic form. One has:

$$\det A = \det \begin{pmatrix} 0 & 2 & -1 & 1 \\ -2 & 0 & -2 & -2 \\ 1 & 2 & 0 & 1 \\ -1 & 2 & -1 & 0 \end{pmatrix} = 4 \neq 0 \qquad (6.0.1)$$

and

$$^t A = -A \qquad (6.0.2)$$

then A is an invertible antisymmetric matrix, thus it defines a symplectic form:

$$\omega_A(X, X') = \langle X, AX' \rangle.$$

Pick

$$f_1 = (0, 0, 0, 1) \quad \text{then} \quad Af_1 = (1, -2, 1, 0). \qquad (6.0.3)$$

One can take $e_1 = (1, 0, 0, 0)$ so that

$$\omega_A(e_1, f_1) = \langle e_1, Af_1 \rangle = 1$$

$V_1 = span\{e_1, f_1\}$ then

$$V_1^{\omega_A} = \left\{ X = (x_1, x_2, x_3, x_4) \text{ s.t} \begin{cases} x_1 & - & 2x_2 & + & x_3 & & & = & 0 \\ & & -2x_2 & + & x_3 & - & x_4 & = & 0 \end{cases} \right\}.$$

Therefore

$$V_1^{\omega_A} = \{v_1 = (-1, 0, 1, 1), v_2 = (0, 1, 2, 0)\} \qquad (6.0.4)$$

and

$$Av_1 = (0, -2, 0, 0), \qquad \omega_A(v_1, v_2) = \langle v_2, Av_1 \rangle = -2.$$

Thus one chooses $e_2 = -\dfrac{1}{2}v_2$ and $f_2 = v_1$ so that

$$(e_1, f_1, e_2, f_2)$$

is a canonical basis. □

Solution 6.2 *(to Exercise 2.2)*

Let $a = (x, \theta) \in T^*N$, $\theta \in T_x^*N$. Consider $\lambda_N \in \Omega^1(T^*N)$ and $\pi :$ $T^*N \longrightarrow N$; $d_a\pi : T_aT^*N \longrightarrow T_{\pi(a)}N \simeq T_xN$. Let $X \in T_aT^*N$ then $(d_a\pi)(X) \in T_xN$

$$\lambda_N(a)\big(X\big) = \big\langle \theta, (d_a\pi)(X) \big\rangle.$$

Let $\big(\mathcal{U}, (x_1, \cdots, x_n)\big)$ be a local coordinates and $\big(T^*\mathcal{U}, (x_1, \cdots, x_n, y_1, \cdots, y_n)\big)$ its corresponding. Hence $\lambda_N|_\mathcal{U} \in T^*\mathcal{U} \subset T^*N$ can be expressed as:

$$\lambda_N|_\mathcal{U} = \sum_{i=1}^n f_i dx_i + g_i dy_i$$

where f_i, g_i are smooth functions on $\mathcal{U}$. Evaluating λ_N on basis vectors $\left(\dfrac{\partial}{\partial x_i}, \dfrac{\partial}{\partial y_i}\right)$ one obtains:

$$\begin{cases} f_i &= \lambda_N|_\mathcal{U}\left(\dfrac{\partial}{\partial x_i}\right) &= \theta\left(\dfrac{\partial}{\partial x_i}\right) &= y_i \\ g_i &= \lambda_N|_\mathcal{U}\left(\dfrac{\partial}{\partial y_i}\right) &= \theta\left(\dfrac{\partial}{\partial y_i}\right) &= 0. \end{cases}$$

Therefore

$$\lambda_N|_\mathcal{U} = \sum_{i=1}^n y_i dx_i. \tag{6.0.5}$$

□

Solution 6.3 *(to Exercise 2.3)*

Let $\alpha \in \Omega^1(N)$ a 1-form on N. View as a section $N \xrightarrow{\alpha} T^*N \xrightarrow{\pi} N$. Let $x \in N$, $X \in T_xN$ and $\alpha(x) = a = (x, \alpha_x) \in T_x^*N$ $d_a\pi : T_aT^*N \longrightarrow T_xN$. One has:

$$\begin{aligned} \big(\alpha^*(\lambda_N)\big)_x(X) &= \lambda_N\big(\alpha(x)\big)\big(d_x\alpha(X)\big) \tag{6.0.6} \\ &= \alpha_x\big(d_a\pi\big(d_x\alpha(X)\big)\big) \\ &= \alpha_x\big(d_x(\alpha \circ \pi)X\big) = \alpha_x(X) \end{aligned}$$

thus

$$\alpha^* \lambda_M = \alpha.$$

$\square$

Solution 6.4 *(to Exercise 2.7)*

Let $\varphi : N_1 \longrightarrow N_2$ be a diffeomorphism between two n-dimensional manifolds N_1 and N_2. Consider $\varphi^{-1} : N_2 \longrightarrow N_1$ and let $\varphi(x) \in N_2$, $a = (x, \theta_x) \in T^* M$. Then

$$d_{\varphi(x)} \varphi^{-1} : T_{\varphi(x)} N_2 \longrightarrow T_x N_1.$$

Let

$$\tilde{\varphi}(a) = \tilde{\varphi}(x, \theta_x) = \left(\varphi(x), \ {}^t\!\left(d_{\varphi(x)} \varphi^{-1} \right) \theta_x \right), \tag{6.0.7}$$

one has

$$\tilde{\varphi}^{-1}(y, \eta_y) = \left(\varphi^{-1}(y), \ {}^t\!\left(d_{\varphi^{-1}(y)} \varphi \right) \eta_y \right). \tag{6.0.8}$$

Let $\xi \in T_a(T^* M)$

$$
\begin{aligned}
(\tilde{\varphi}^* \lambda_N)(x, \theta_x)\left(\xi \right) &= \lambda_N\!\left(\tilde{\varphi}(x, \theta_x) \right)\left((d_{(x,\theta_x)} \tilde{\varphi}) \xi \right) \tag{6.0.9} \\[2mm]
&= \ {}^t\!\left(d_{\varphi(x)} \varphi^{-1} \right) \theta_x \left(d_{\tilde{\varphi}(x,\theta_x)} \pi_N \left(d_{(x,\theta_x)} \tilde{\varphi} \right) \xi \right) \\[2mm]
&= \left\langle \theta_x, d_{\varphi(x)} \varphi^{-1} \left(d_{\tilde{\varphi}(x,\theta_x)} \pi_N \left(d_{(x,\theta_x)} \tilde{\varphi} \right) \xi \right) \right\rangle \\[2mm]
&= \left\langle \theta_x, d_{\varphi(x)} \varphi^{-1} \left(d_{(x,\theta_x)} (\pi_N \circ \tilde{\varphi}) \xi \right) \right\rangle \\[2mm]
&= \left\langle \theta_x, d_{\varphi(x)} \varphi^{-1} \left(d_{(x,\theta_x)} (\varphi \circ \pi_M) \xi \right) \right\rangle \\[2mm]
&= \left\langle \theta_x, d_x (\varphi^{-1} \circ \varphi) \left(d_{(x,\theta_x)} \pi_M \xi \right) \right\rangle \\[2mm]
&= \left\langle \theta_x, \left(d_{(x,\theta_x)} \pi_M \xi \right) \right\rangle = \lambda_M(x, \theta_x)\left(\xi \right).
\end{aligned}
$$

Consequently $\tilde{\varphi}$ is symplectic since

$$d(\tilde{\varphi}^*\lambda_M) = \tilde{\varphi}^*(d\lambda_N).$$

$\square$

Solution 6.5 *(to Exercise 3.3)*

Let $f, g, h \in \mathcal{C}^\infty(M, \omega)$ and $\alpha, \beta \in \mathbb{R}$.

1. It is direct from the fact that ω is skew symmetric:

$$\{f, g\} = \omega(X_f, X_g) = -\omega(X_g, X_f) = -\{g, f\}.$$

2.
$$
\begin{aligned}
\{(\alpha f + \beta g), h\} &= \omega(X_{(\alpha f + \beta g)}, X_h) \\
&= \omega(X_{\alpha f}, X_h) + \omega(X_{\beta g}, X_h) \\
&= \alpha\omega(X_f, X_h) + \beta\omega(X_g, X_h) \\
&= \alpha\{f, h\} + \beta\{g, h\}. \qquad (6.0.10)
\end{aligned}
$$

Thus $\{\cdot, \cdot\}$ is linear in the first entry, so that it is bilinear since it is skew-symmetric.

3. We know that for all $X, Y, Z \in \mathfrak{X}(M)$:

$$\big[X, [Y, Z]\big] + \big[Y, [Z, X]\big] + \big[Z, [X, Y]\big] = 0,$$

$$i_{[X,Y]} = i_X L_Y - L_Y i_X$$

and

$$L_{[X,Y]} = -[L_X, L_Y].$$

Thus

$$
\begin{aligned}
i_{\left[X_f,[X_g,X_h]\right]}\omega &= i_{X_f}L_{[X_g,X_h]}\omega - L_{[X_g,X_h]}i_{X_f}\omega \\
&= -i_{X_f}\left(L_{X_g}L_{X_h} - L_{X_h}L_{X_g}\right)\omega \\
&\quad +\left(L_{X_g}L_{X_h} - L_{X_h}L_{X_g}\right)i_{X_f}\omega \\
&= \left(L_{X_g}L_{X_h} - L_{X_h}L_{X_g}\right)i_{X_f}\omega \\
&= L_{X_g}\left(i_{X_h}di_{X_f} + di_{X_h}i_{X_f}\right)\omega \\
&\quad -L_{X_h}\left(i_{X_g}di_{X_f} + di_{X_g}i_{X_f}\right)\omega \\
&= L_{X_g}\left(di_{X_h}i_{X_f}\right)\omega - L_{X_h}\left(di_{X_g}i_{X_f}\right)\omega \\
&= d\left(L_{X_g}\omega(X_h,X_f)\right) - d\left(L_{X_h}\omega(X_g,X_f)\right) \\
&= d\left(i_{X_g}d\{h,f\}\right) - d\left(i_{X_h}d\{g,f\}\right) \\
&= d\{\{h,f\},g\} - d\{\{g,f\},h\} \\
&= d\{g,\{f,h\}\} - d\{h,\{f,g\}\}. \tag{6.0.11}
\end{aligned}
$$

In the same manner we have:

$$
i_{\left[X_g,[X_h,X_f]\right]}\omega = d\{h,\{g,f\}\} - d\{f,\{g,h\}\}, \tag{6.0.12}
$$

$$
i_{\left[X_h,[X_f,X_g]\right]}\omega = d\{f,\{h,g\}\} - d\{g,\{h,f\}\}. \tag{6.0.13}
$$

Thus one obtains:

$$
\begin{aligned}
\kappa &= i_{\left[X_f,[X_g,X_h]\right]+\left[X_g,[X_h,X_f]\right]+\left[X_h,[X_f,X_g]\right]}\omega \\
&= 2d\Big(\{g,\{f,h\}\} + \{h,\{g,f\}\} + \{f,\{h,g\}\}\Big) \\
&= 0 \qquad \forall\, f,g,h \in C^\infty(M), \tag{6.0.14}
\end{aligned}
$$

which infers

$$
\{g,\{f,h\}\} + \{h,\{g,f\}\} + \{f,\{h,g\}\} = 0 \ \forall\, f,g,h \in C^\infty(M).
$$

4.
$$
\begin{aligned}
\{f, u\cdot v\} &= -d(u\cdot v)(X_f) = -\Big(v(du) + u(dv)\Big)(X_f) \\
&= -vdu(X_f) - udv(X_f) \\
&= -v\{u,f\} - u\{v,f\} = v\{f,u\} + u\{f,v\}.
\end{aligned}
$$

$\square$

Solution 6.6 *(to Exercise 5.3)*

The map $\varphi : \mathbb{R}^3 \longrightarrow \mathbb{R}^3$,

$$(x, y, z) \longrightarrow (z \cos x - y \sin x, -z \sin x - y \cos x, -x)$$

pulls back $(\cos z)dx + (\sin z)dy$ to $dz - xdy$. □

Solution 6.7 *(to Exercise 5.5)*

Let $(x, y, z) \in \mathbb{R}^3$. One has:

$$(x, y, z) \perp (0, 0, 0) = \big(x + 0, y + 0, z + 0 + (x \cdot 0 - y \cdot 0)\big) = (x, y, z)$$

saying $(0, 0, 0)$ is the zero of $(\mathbb{R}^3, \perp)$. Also

$$(x, y, z) \perp (-x, -y - z) = \big(x - x, y - y, z - z + (-xy + yx)\big) = (0, 0, 0)$$

meaning $(-x, -y, -z)$ is the symmetric of (x, y, z). Let (x, y, z); (u, v, w); (α, β, γ).

Associativity: Let $X \in R^3$ equals to $\big[(x, y, z) \perp (u, v, w)\big] \perp (\alpha, \beta, \gamma)$. We have

$$
\begin{aligned}
X &= \big[(x, y, z) \perp (u, v, w)\big] \perp (\alpha, \beta, \gamma) \\
&= \Big(x + u + \alpha, y + v + \beta, z + w + \gamma + (x \cdot v - y \cdot u) \\
&\quad + \big(\beta \cdot (x + u) - \alpha \cdot (y + v)\big)\Big) \\
&= \Big(x + u + \alpha, y + v + \beta, z + w + \gamma + (u \cdot \beta - v \cdot \alpha) \\
&\quad + \big(x \cdot (v + \beta) - y \cdot (u + \alpha)\big)\Big) \\
&= (x, y, z) \perp \big[(u, v, w) \perp (\alpha, \beta, \gamma)\big].
\end{aligned}
$$

Let $a = (a_1, a_2, a_3) \in \Big(\mathbb{R}^3, \beta_1 = \sum\limits_{i=1}^{n}(x_i dy_i - y_i dx_i) + dz\Big)$ and $R_a(x, y, z) = (x, y, z) \perp a = (x', y', z')$ be the right translation.

$$
\begin{aligned}
R_a^* \beta_1 &= \sum_{i=1}^{n}(x_i' dy_i' - y_i' dx_i') + dz' \\
&= \sum_{i=1}^{n}\Big((x_i + a_1)d(y_i + a_2) - (y_i + a_2)d(x_i + a_1)\Big) + d(z + a_3) \\
&= \sum_{i=1}^{n}(x_i dy_i - y_i dx_i) + dz = \beta_1. \tag{6.0.15}
\end{aligned}
$$

Solution 6.8 *(to Exercise 5.8)*

Let α be a contact form on a differential manifold M and λ a positive smooth function on M. Consider $\alpha_\lambda = \lambda\alpha$.

$$d\alpha_\lambda|_E = \lambda\Big(d\alpha|_E\Big).$$

Let X be the Reeb field of α_λ, i.e

$$\begin{cases} i_X\alpha_\lambda & = & 1 \qquad (1) \\ i_X d\alpha_\lambda & = & 0 \qquad (2). \end{cases} \qquad (6.0.16)$$

$(1) \Longrightarrow i_X\alpha_\lambda = \lambda\alpha(X) = 1$ then $X_0 = \frac{1}{\lambda}\xi$ is a good candidate. Let us now plug X_0 in (2), we have

$$\begin{aligned} i_{X_0}d\alpha_\lambda & = & i_{(\frac{1}{\lambda}\xi)}\Big(d\lambda \wedge \alpha + \lambda d\alpha\Big) \\ & = & \Big(\frac{1}{\lambda}d\lambda(\xi)\Big)\alpha - \frac{1}{\lambda}d\lambda \qquad (6.0.17) \end{aligned}$$

which is semi-basic so it rises from a horizontal vectors field (that is a section of E)

$$\begin{aligned} i_{X_0}d\alpha_\lambda & = & \Big(d\ln\lambda)\xi\Big)\alpha - d(\ln\lambda) \\ & = & \widetilde{(d\alpha)}(H_{\ln\lambda}) \\ & = & i_{H_{\ln\lambda}}d\alpha = \frac{1}{\lambda}i_{H_{\ln\lambda}}d\alpha_\lambda = i_{\left(\frac{1}{\lambda}H_{\ln\lambda}\right)}d\alpha_\lambda \qquad (6.0.18) \end{aligned}$$

thus

$$i_{X_0}d\alpha_\lambda - i_{\left(\frac{1}{\lambda}H_{\ln\lambda}\right)}d\alpha_\lambda = 0 \Leftrightarrow i_{(X_0 - \frac{1}{\lambda}H_{\ln\lambda})}d\alpha_\lambda = 0; \qquad (6.0.19)$$

morever

$$\alpha_\lambda\left(\frac{1}{\lambda}(\xi - H_{\ln\lambda})\right) = \alpha_\lambda\left(\frac{1}{\lambda}\xi\right) = 1.$$

Therefore

$$X = \frac{1}{\lambda}\Big(\xi - H_{\ln\lambda}\Big)$$

is the Reeb field of α_λ.   □

Solution 6.9 *(to Exercise 5.9)*

Explicit formula for the bracket $[f, g]$ for two functions $f, g \in C^{\infty}(R^{2n+1})$, where R^{2n+1} is equipped with $\alpha = \sum_{i=1}^{n} y_i dx_i + dz$. Let ξ be the Reeb field of α.

Let $f, g \in C^{\infty}(R^{2n+1})$. The bracket is given by: $[f, g] = \alpha([Y_f, Y_g])$ where $Y_f = H_f + f\xi$, $H_f = \widetilde{d\alpha}(\theta_f)$, $\theta_f = (\xi \cdot f)\alpha - df$ and similarly $Y_g = H_g + g\xi$, $H_g = \widetilde{d\alpha}(\theta_g)$, and $\theta_g = (\xi \cdot g)\alpha - dg$. One has:

$$\alpha = \sum_{i=1}^{n} y_i dx_i + dz \implies \xi = \frac{\partial}{\partial z} \text{ and } d\alpha = dy_i \wedge dx_i. \qquad (6.0.20)$$

Thus:

$$\theta_f = \left(y_i \frac{\partial f}{\partial z} - \frac{\partial f}{\partial x_i} \right) dx_i - \frac{\partial f}{\partial y_i} dy_i,$$

$$H_f = \frac{\partial f}{\partial y_i} \frac{\partial}{\partial x_i} + \left(y_i \frac{\partial f}{\partial z} - \frac{\partial f}{\partial x_i} \right) \frac{\partial}{\partial y_i}, \qquad (6.0.21)$$

$$H_f = \frac{\partial f}{\partial y_i} \frac{\partial}{\partial x_i} + \left(y_i \frac{\partial f}{\partial z} - \frac{\partial f}{\partial x_i} \right) \frac{\partial}{\partial y_i} + f \frac{\partial}{\partial z};$$

and similarly

$$\theta_g = \left(y_i \frac{\partial g}{\partial z} - \frac{\partial g}{\partial x_i} \right) dx_i - \frac{\partial g}{\partial y_i} dy_i,$$

$$H_g = \frac{\partial g}{\partial y_i} \frac{\partial}{\partial x_i} + \left(y_i \frac{\partial g}{\partial z} - \frac{\partial g}{\partial x_i} \right) \frac{\partial}{\partial y_i}, \qquad (6.0.22)$$

$$H_g = \frac{\partial g}{\partial y_i} \frac{\partial}{\partial x_i} + \left(y_i \frac{\partial g}{\partial z} - \frac{\partial g}{\partial x_i} \right) \frac{\partial}{\partial y_i} + g \frac{\partial}{\partial z}.$$

Therefore

$$
\begin{aligned}
[Y_f, Y_g] &= \left(f\frac{\partial^2 g}{\partial y_i \partial z} + \left(y_i\frac{\partial f}{\partial z} - \frac{\partial f}{\partial x_i} \right)\frac{\partial^2 g}{\partial y_i^2} \right. \\
&\quad \left. -g\frac{\partial^2 f}{\partial y_i \partial z} - \left(y_i\frac{\partial g}{\partial z} - \frac{\partial g}{\partial x_i} \right)\frac{\partial^2 f}{\partial y_i^2} \right)\frac{\partial}{\partial x_i} \\
&\quad + \left(f\left(y_i\frac{\partial^2 g}{\partial z^2} - \frac{\partial^2 g}{\partial x_i \partial z} \right) + \frac{\partial f}{\partial y_i}\left(y_i\frac{\partial^2 g}{\partial x_i \partial z} - \frac{\partial^2 g}{\partial x_i^2} \right) \right. \\
&\quad \left. -g\left(y_i\frac{\partial^2 f}{\partial z^2} - \frac{\partial^2 f}{\partial x_i \partial z} \right) - \frac{\partial g}{\partial y_i}\left(y_i\frac{\partial^2 f}{\partial x_i \partial z} - \frac{\partial^2 f}{\partial x_i^2} \right) \right)\frac{\partial}{\partial y_i} \\
&\quad + \left(\frac{\partial f}{\partial y_i}\frac{\partial q}{\partial x_i} + \left(y_i\frac{\partial f}{\partial z} - \frac{\partial f}{\partial x_i} \right)\frac{\partial g}{\partial y_i} - \frac{\partial g}{\partial y_i}\frac{\partial f}{\partial x_i} - \left(y_i\frac{\partial g}{\partial z} - \frac{\partial g}{\partial x_i} \right) \right)\frac{\partial}{\partial z} \\
&\quad + \left(\frac{\partial f}{\partial y_i}\left(y_i\frac{\partial g}{\partial z} - \frac{\partial g}{\partial x_i} \right) - \frac{\partial g}{\partial y_i}\left(y_i\frac{\partial f}{\partial z} - \frac{\partial f}{\partial x_i} \right) \right)\left[\frac{\partial}{\partial x_i}, \frac{\partial}{\partial y_i} \right] \\
&\quad + \left(g\frac{\partial f}{\partial y_i} - f\frac{\partial g}{\partial y_i} \right)\left[\frac{\partial}{\partial x_i}, \frac{\partial}{\partial z} \right] \\
&\quad + \left(g\left(y_i\frac{\partial f}{\partial z} - \frac{\partial f}{\partial x_i} \right) - f\left(y_i\frac{\partial g}{\partial z} - \frac{\partial g}{\partial x_i} \right) \right)\left[\frac{\partial}{\partial y_i}, \frac{\partial}{\partial z} \right].
\end{aligned}
$$

Using the following relations

$$
\alpha\left(\frac{\partial}{\partial x_i} \right) = y_i, \qquad \alpha\left(\frac{\partial}{\partial y_i} \right) = 0, \qquad \alpha\left(\frac{\partial}{\partial z} \right) = 1,
$$

$$
\alpha\left(\left[\frac{\partial}{\partial x_i}, \frac{\partial}{\partial y_i} \right] \right) = \frac{\partial}{\partial x_i}\alpha\left(\frac{\partial}{\partial y_i} \right) - \frac{\partial}{\partial y_i}\alpha\left(\frac{\partial}{\partial x_i} \right) - d\alpha\left(\frac{\partial}{\partial x_i}, \frac{\partial}{\partial y_i} \right) = 0,
$$

$$
\alpha\left(\left[\frac{\partial}{\partial x_i}, \frac{\partial}{\partial z} \right] \right) = \frac{\partial}{\partial x_i}\alpha\left(\frac{\partial}{\partial z} \right) - \frac{\partial}{\partial z}\alpha\left(\frac{\partial}{\partial x_i} \right) - d\alpha\left(\frac{\partial}{\partial x_i}, \frac{\partial}{\partial z} \right) = 0,
$$

$$
\alpha\left(\left[\frac{\partial}{\partial y_i}, \frac{\partial}{\partial z} \right] \right) = \frac{\partial}{\partial y_i}\alpha\left(\frac{\partial}{\partial z} \right) - \frac{\partial}{\partial z}\alpha\left(\frac{\partial}{\partial y_i} \right) - d\alpha\left(\frac{\partial}{\partial y_i}, \frac{\partial}{\partial z} \right) = 0,
$$

one finally obtains

$$\begin{aligned}
\left[f,g\right] \;=\; & y_i \left(f \frac{\partial^2 g}{\partial y_i \partial z} + \left(y_i \frac{\partial f}{\partial z} - \frac{\partial f}{\partial x_i} \right) \frac{\partial^2 g}{\partial y_i^2} - g \frac{\partial^2 f}{\partial y_i \partial z} \right. \\
& \left. - \left(y_i \frac{\partial g}{\partial z} - \frac{\partial g}{\partial x_i} \right) \frac{\partial^2 f}{\partial y_i^2} \right) \\
& + \left(\frac{\partial f}{\partial y_i} \frac{\partial g}{\partial x_i} + \left(y_i \frac{\partial f}{\partial z} - \frac{\partial f}{\partial x_i} \right) \frac{\partial g}{\partial y_i} - \frac{\partial g}{\partial y_i} \frac{\partial f}{\partial x_i} \right. \\
& \left. - \left(y_i \frac{\partial g}{\partial z} - \frac{\partial g}{\partial x_i} \right) \right).
\end{aligned}$$

$\square$

Chapter 7

Epilogue: The C^0-symplectic and contact topology

Symplectic and Contact Geometry belong to the C^∞ category: *the objects are smooth manifolds equipped with smooth differential forms (symplectic forms or contact forms) and the morphisms are smooth diffeomorphisms preserving the smooth structures defined by these forms.* The natural topology to work with is the C^∞ compact-open topology (see [Hir76]).

A quite new discipline, called the C^0 *symplectic and contact topology* examines the interplay of the symplectic/contact objects with the (C^0) uniform topology [Hum08].

The typical question is the following:

Given a sequence (S_n) of symplectic/contact objects, which converge uniformly to some object S, did the symplectic/contact nature of the objects S_n survive the passage through the uniform limit?

Amazingly enough, there are some "wonders" in Symplectic Geometry and Contact Geometry in which the symplectic/contact nature survive!

The first question raised was the following:

Given a sequence ϕ_n of symplectic diffeomorphisms of a symplectic manifold (M,ω), which converges uniformly to a smooth diffeomorphism ϕ, is ϕ a symplectic diffeomorphism?

The question can also be formulated this way:

Is the group $\mathrm{Symp}(M,\omega)$ of symplectic diffeomorphisms of (M,ω) C^0-closed in the group $\mathrm{Diff}^\infty(M)$ of all smooth diffeomorphisms of M?

A wonder of Symplectic Geometry is that the answer to this question is yes.

Theorem 7.1 *(Eliashberg-Gromov rigidity theorem)*

If a sequence $\phi_n \in Symp(M, \Omega)$ converges uniformly to a diffeomorphism ϕ then $\phi \in Symp(M, \Omega)$.

This theorem was proved by Gromov in [Gro86] and independently by Eliashberg in [Eli87].

Here we give a short and very clever proof found recently by Buhovsky [Buh14].

Another fascinating result is the following:

Theorem 7.2

Let Φ_{H_n}, Φ_{K_n} be two sequences of hamiltonian diffeomorphisms, with Hamiltonians H_n and K_n respectively. Suppose that Φ_{H_n} (resp. Φ_{K_n}) converges uniformly to homeomorphisms ϕ (resp. ψ) and H_n (resp. K_n) converges to continuous functions H (resp. K). Then $\phi = \psi$ if and only if $H = K$.

The fact that $H = K$ implies that $\phi = \psi$ is due to Hofer-Zehnder [Hof-Zeh94] and Oh-Muller [Oh-Mül07].

The fact that $\phi = \psi$ implies that $H = K$, was first proved by Viterbo [Vit06] . Buhovsky-Seyfaddini subsequently [Buh-Sey13] found a generalization of Viterbo result where it is sufficient for the sequences H_n and K_n of normalized hamiltonians to converge in the Hofer norm.

We insist that Theorem 7.2 is not trivial: because we cannot use the theorem of existence and uniqueness of solutions of ODE.

7.1   The Hofer norm [Hof90]

Let us now recall the Hofer norm on $Ham(M, \omega)$: an isotopy of a smooth manifold M is a smooth family $\Phi = \phi_t$ of diffeomorphisms $\phi_t : M \longrightarrow M$ with $\phi_0 = id$. For each isotopy (ϕ_t), we consider the family of vector field $\dot{\phi}_t$ defined by:

$$\dot{\phi}_t(x) = \frac{d\phi_t}{dt}\big(\phi_t^{-1}(x)\big).$$

Let (M, ω) be a symplectic manifold. A Hamiltonian isotopy of (M, ω) with compact support is an isotopy (ϕ_t) with compact support such that there exists a smooth family $H = (H_t)$ of smooth functions with compact support on (M, ω) such that $i_{\dot{\phi}_t}\omega = -dH_t$. We will denote such isotopy by $\Phi_H = (\phi_t)$, where $i_\xi\omega(X) = \omega(\xi, X)$ for all vector field X.

The set of time-one maps of Hamiltonian isotopies with compact supports forms a group denoted $Ham(M, \omega)$ and called the group of **hamiltonian diffeomorphisms** of (M, ω), [Ban97].

The group $Ham(M, \omega)$ carries a celebrated bi-invariant metric discovered by Hofer [Hof90]. For $\Phi \in Ham(M, \omega)$, Hofer defined:

$$\|\phi\| := \inf_{\Phi_H} \int_0^1 \Big(\max_x \big(H(x, t)\big) - \min_x \big(H(x, t)\big) \Big) dt$$

where the infimum is taken over all Hamiltonian isotopy Φ_H having ϕ as the time-one map. The expression

$$l(\Phi) := \int_0^1 \Big(\max_x (H(x, t)) - \min_x (H(x, t)) \Big) dt$$

is called the Hofer length of the isotopy Φ.

The oscillation $osc(f)$ of a function with compact support is:

$$osc(f) = \max_x f(x) - \min_x f(x).$$

The Hofer norm of ϕ is defined as

$$\|\phi\| = \inf_{\Phi_H} \int_0^1 osc\big(H(x, t)\big) dt.$$

It is easy to show that this is a pseudo-metric; however it is very difficult to show that it is non-degenerate. This was proved in its full generality by Lalonde-McDuff [Lal-McD95].

The Hofer distance d_H between two Hamiltonian diffeomorphisms ϕ and ψ is defined as

$$d_H(\phi, \psi) = \|\phi \cdot \psi\|.$$

Generalization of Hofer norm [Ban10] and the set of string symplectic homeomorphisms.

Let $Iso(M, \omega)$ denote the space of symplectic isotopies of a closed symplectic manifold (M, ω), i.e the set of smooth maps $\Phi : M \times [0, 1] \longrightarrow M$ such that for all $t \in [0, 1]$, $\phi_t : M \longrightarrow M$, $x \longmapsto \Phi(x, t)$ is a symplectic diffeomorphism and $\phi_0 = id$ and denote by $\text{Symp}(M, \omega)$ the group of symplectic diffeomorphisms isotopic to the identity, i.e the time-one maps of elements in $Iso(M, \omega)$.

The "Lie algebra" of $\mathrm{Symp}(M, \omega)$ is the space $\mathrm{symp}(M, \omega)$ of symplectic vector fields, i.e the set of vector fields X such that $i_X \omega$ is a closed form.

When M is a compact manifold, we define a norm $\| \cdot \|$ on $\mathrm{symp}(M, \omega)$ as follows: first we fix a riemannian metric g. For any $X \in \mathrm{symp}(M, \omega)$, we consider the Hodge decomposition of $i_X \omega$: there is a unique harmonic 1-form $\mathcal{H}_X$ and a unique function u_X such that

$$i_X \omega = \mathcal{H}_X + du_X.$$

This defines a decomposition of $X \in \mathrm{symp}(M, \omega)$ as:

$$X = \#H_X + X_{u_X}$$

where $\#H_X$ is defined by the equation $i_{\#H_X} \omega = \mathcal{H}_x$ and X_{u_X} is the Hamiltonian vector field with u_X as hamiltonian.

Recall that the space $harm^1(M, g)$ of harmonic 1-forms is a finite dimensional vector space over $\mathbb{R}$ and its dimension is the first Betti number b_1 of M. On $harm^1(M, g)$ we put the following "Euclidean" norm: for $\mathcal{H} \in harm^1(M, \omega)$, $\mathcal{H} = \sum_i \lambda_i h_i$ define $|\mathcal{H}|_{\mathcal{B}} := \sum_i |\lambda_i|$ where $\mathcal{B} = (h_1, h_2, \cdots, h_{b_1})$ is a basis of $harm^1(M, g)$.

Now we define the norm $\| \cdot \|$ on the vector space $\mathrm{symp}(M, \omega)$ by:

$$\|X\| = |\mathcal{H}_X|_{\mathcal{B}} + osc(u_X) \tag{7.1.1}$$

where $|\mathcal{H}_X|_{\mathcal{B}}$ is the euclidean norm of the harmonic 1 form $\mathcal{H}_X$.

Remark 7.1
   This norm is equivalent to the restriction to $harm^1(M, g)$ of the l^2–norm on p-forms:

$$\|\alpha\| = \int_M \alpha \wedge *\alpha$$

where the $*$ is the Hodge star operator of the riemannian metric g.

Theorem 7.3 *[Ban10]*
   The topology defined by the metric above on $symp(M, \omega)$ is independent of the choice of the riemannian metric g.

Let φ_t be a symplectic isotopy, then

$$\dot{\phi}_t(x) = \frac{d\phi_t}{dt}\left(\phi_t^{-1}(x)\right)$$

is a smooth family of symplectic vector field. We defined the length of a symplectic isotopy $\Phi = (\phi_t)$:

$$l(\Phi) = \frac{1}{2}\left(l_0(\Phi) + l_0(\Phi^{-1})\right)$$

where

$$l_0(\Phi) := \int_0^1 \left(|\mathcal{H}_\Phi| + osc(u_\Phi)\right)dt$$

and Φ^{-1} is the isoptopy $\{\phi_t^{-1}\}$.

Here H_Φ and u_Φ come from the Hodge decomposition

$$i_{\dot{\phi}_t}\omega = H_\Phi + d(u_\phi).$$

We denoted by $|H_\Phi|$ the norm defined on the finite dimensional vector space of harmonic 1-forms.

Given $\phi \in Symp(M,\omega)$, we define the *Hofer-like metric* $\|\cdot\|_{HL}$ by:

$$\|\phi\|_{HL} = \inf l(\phi)$$

where the infimum is taken over all symplectic isotopies $\Phi = \{\phi_t\}$, with $\phi_1 = \phi$.

Remark 7.2

If $\{\phi_t\}$ is a symplectic isotopy, we identify it with (H_Φ, u_Φ), or simply (H, u) when $i_{\dot{\phi}_t}\omega = H_\Phi + du_\Phi$.

Theorem 7.4 *[Ban10]*

$\|\cdot\|_{HL}$ *is a norm on* $Symp(M,\omega)$, *which generalizes the Hofer norm on* $Ham(M,\omega)$.

For Φ, $\Psi \in Iso(M,\omega)$, we define the distance $D(\Phi, \Psi)$ by:

$$D(\Phi, \Psi) = \int_0^1 \|\dot{\phi}_t - \dot{\psi}_t\|dt.$$

The C^0 topology on the group $Homeo(M)$ of all compactly supported homeomorphism of a smooth manifold M coincides with the metric topology coming from the metric

$$\bar{d}(g,h) = \max \left(\sup_{x \in M} d_0\big(g(x), h(x)\big), \sup_{x \in M} d_0\big(g^{-1}(x), h^{-1}(x)\big)\right)$$

where d_0 is a distance on M induced by some riemannian metric. On the space $PHomeo(M)$ of continuous paths $\gamma : [0,1] \longrightarrow Homeo(M)$ one has the distance

$$\bar{d}(\gamma, \mu) = \sup_{t \in [0,1]} \bar{d}\big(\gamma(t), \mu(t)\big).$$

Definition 7.1

The symplectic topology on $Iso(M, \omega)$ is the topology induced by the symplectic distance:

$$d_{symp} = \bar{d} + D.$$

The topology is independent of the choices made in defining the symplectic distance.

The symplectic distance is a generalization of the Hamiltonian distance d_{ham} of Oh-Müller [Oh-Mül07] who used it to define the notion of "Hamiltonian homeomorphisms".

The following result is due to Banyaga-Hurtubise-Speach and Tchuiaga.

Theorem 7.5

Let Φ_n be a sequence of symplectic isotopies and Ψ another symplectic isotopy. Suppose the time-one map of Φ_n converges uniformly to some hoimeomorphism ϕ. If $l(\Phi_n \Psi^{-1})$ converges to 0 as n goes to ∞, then $\phi = \Psi_1$, where Ψ_1 is the time-one map of Ψ.

The following uniqueness theorem is due to Banyaga-Tchuiaga [Ban-Tch14].

Theorem 7.6

Let $\Phi_n = (\phi_n^t) = (H_n^t, u_n^t)$ and $\Psi_n = (\psi_n^t) = (H_n'^t, u_n'^t)$ be two sequences of symplectic isotopies, which converge uniformly to the same limit of homeomorphisms, and H_n^t, $H_n' \longrightarrow H^t$, H'^t in the l^2 norm and u_n^t, $u_n' \longrightarrow u^t$, u'^t in the Hofer norm, then $H^t = H'^t$ and $u^t = u'^t$.

This theorem justifies the following:

Definition 7.2

A homeomorphism $h : M \longrightarrow M$ is called a strong symplectic homem-orphism if there exists a d_{symp}-Cauchy sequence $\Phi^n = (\phi_t^n)$ of symplectic isotopies such that ϕ_t^n converges uniformly to h.

Theorem 7.7

The set $SSympeo(M, \omega)$ of all strong symplectic homeomorphisms forms a group.

Proof of Eliashberg-Gromov rigidity theorem:

We reproduce here beautiful argument of Buhovsky [Buh14].

Let ϕ_n be a sequence of symplectic diffeomorphisms converging uniformly to a diffeomorphism ϕ.

If ϕ is not symplectic, there exists a smooth function $F : M \longrightarrow \mathbb{R}$ such that

$$\phi_*(X_F) \neq X_{F \circ \phi}. \tag{7.1.2}$$

(see Theorem 3.1).

Let Φ_u be the flow of X_u.

The equation (7.1.2) implies that

$$\phi \Phi_F \phi^{-1} \neq \Phi_{F \circ \phi}.$$

Consider now the sequences of Hamiltonian isotopies:

$$\sigma_n = \phi_n \Phi_F \phi_n^{-1}$$

with Hamiltonian $F \circ \phi_n$ which converge uniformly (hence in the $L^{(1,\infty)}$ norm) to $F \circ \phi$ and the constant sequence

$$\mu_n = \phi_{F \circ \phi}$$

for all n, with hamiltonian $F \circ \phi$ converging to $F \circ \phi$.

Since the Hamiltonians of σ and μ_n converge to the same function, Oh-Müller uniqueness theorem then asserts that, σ_n and μ_n have the same limit, i.e

$$\varphi \Phi_F \varphi^{-1} = \Phi_F \circ \varphi$$

which implies that

$$\phi_*(X_F) = X_{F \circ \phi}$$

contradicting the equation (7.1.2). $\square$

Displacement energy

In [Hof90], Hofer introduced the notion of displacement energy $e(A)$ of a bounded subset $A \subset M$:

$$e(A) = \inf\left\{\|\phi\|_H, \phi \in Ham(M,\omega) \ \phi(A) \cap A = \emptyset\right\}.$$

We have the following result:

Theorem 7.8 *(Eliashberg-Polterovich)[Eli-Pol93]*
    For any non-empty open subset A of M, $e(A)$ is a strictly positive number.

The following notion was proposed in [Ban-Hur-Spa16].

Definition 7.3
    The symplectic displacement energy $e_S(A)$ of a subset A of M is:

$$e_S(A) = \inf\left\{\|\phi\|_{HL}, \phi \in Symp(M,\omega) \ \phi(A) \cap A = \emptyset\right\}.$$

Theorem 7.9 *(Banyaga-Hurtubise-Spaeth)*
    For any non-empty open set A, $e_S(A)$ is a strictly positive number.

Proof of Theorem 7.5
    Suppose $\phi \neq \Psi_1$, i.e $\phi^{-1}\Psi_1 \neq id$. There exists a small ball B such that $\phi^{-1}\Psi_1(B) \cap B = \emptyset$. Since Φ_n^1 converges uniformly to ψ, $\left((\Phi_n^1)^{-1}\Psi(B)\right) \cap B = \emptyset$ for n large enough. If $e_S(B)$ is the symplectic energy of B, then

$$e_S(B) \leqslant \|(\Phi_n^1)^{-1}\Psi\|_{HL} \leqslant l\left((\Phi_n^1)^{-1}\Psi\right) \longrightarrow 0$$

as $n \longrightarrow +\infty$. This contradicts the positivity of $e_S(B)$. $\qquad\square$

7.2 Contact rigidity

Let (M,α) be a contact manifold. Recall that a contactomorphism (or a contact diffeomorphism) is a diffeomorphism $\varphi : M \longrightarrow M$ such that $\varphi^*\alpha = e^h\alpha$ for some smooth function h on M. The function h is called **"the conformal factor"**. If $h = 0$, we say that φ is a **strictly contact**

diffeomorphism: $\varphi^*\alpha = \alpha$.

Denote by $\mathrm{Diff}(M, \alpha)$ resp. $\mathrm{Diff}_\alpha(M)$ the group of all contact diffeomorphisms resp. strictly conatct diffeomorphisms equipped with the C^∞-compact open topology.

We have the following C^0-rigidity theorems:

Theorem 7.10 *(Muller-Spaeth)[Mül-Spa15, Mül-Spa14, Mül-Spa]*
  The group $\mathrm{Diff}(M, \alpha)$ *is* C^0-*closed in* $\mathrm{Diff}^\infty(M)$.

Theorem 7.11 *(Banyaga-Spaeth)[Ban-Spa]*
  The group $\mathrm{Diff}_\alpha(M, \alpha)$ *is* C^0-*closed in* $\mathrm{Diff}^\infty(M)$.

The contact topology on the space of contact isotopies.

Let $\Phi = (\varphi_t)$ be a **contact isotopy**, i.e $\dot{\varphi}_t \in \mathcal{L}(M, \alpha)$. Then $f_t = i_{\dot{\varphi}_t}\alpha$ is the generator of Φ. Namely,

$$Y_{f_t} = \dot{\varphi}_t.$$

Conversely any smooth family of functions $\big(C^\infty(M \times \mathbb{R})\big)$ gives rise to a contact isotopy. On $\big(C^\infty(M \times \mathbb{R})\big)$, Müller and Spaeth have put the following norm: for $F \in \big(C^\infty(M \times \mathbb{R})\big)$, $F(x, t) = f_t(x)$

$$\|F\| = \int_0^1 \left[\mathrm{osc}(f_t) + \frac{1}{\mathrm{Vol}(M)} \int_M f_t\big(\alpha \wedge (d\alpha)^n\big) + h_t(x) \right] dt \qquad (7.2.1)$$

where h_t is the conformal factor of the isotopy generated by Y_{f_t}.

When $h_t = 0$, this is the Banyaga-Donato [Ban-Ine95] norm on strictly contact isotopies.

Definition 7.4
  A **contact homeomorphism** $h : M \longrightarrow M$ *is a homeomorphism which is a uniform limit of time 1 maps* φ_1^n *of a sequence of contact isotopies* $\Phi^n = (\varphi_t^n)$ *whose generators form a Cauchy sequence in the contact norm above.*

Let us denote by $l(\Phi^n)$ the limit of the sequences of generators of Φ^n. We have the following facts:

Theorem 7.12

Let Φ^n and Ψ^n be two sequences of contact isotopies whose time-one maps converge uniformly to h and g and their generators converge (in the contact norm) to $l(\Phi^n)$ and $l(\Psi^n)$. Then

$$h = g \Longleftrightarrow l(\Phi^n)) = l(\Psi^n). \tag{7.2.2}$$

This theorem was proved by Müller-Spaeth [Mül-Spa15, Mül-Spa14, Mül-Spa] and its particular case for strictly contact isotopies by Banyaga-Spaeth [Ban-Spa].

A consequence of Theorem 7.12, we see that a contact homeomorphism h determines a unique "generator" $l(h)$.

For rigidity of the Poisson bracket, we refer to [Pol-Dan14].

Review of calculus on manifolds

This chapter is a review of basis notions in differential geometry. Our general references are [Gui-Ste90], [Ste64] and [War71].

A.1 Differential forms and de Rham cohomology

Let M be a smooth n-dimensional manifold, TM its tangent bundle: $TM = \bigcup_{x \in M} T_x M$, where $T_x M$ is the tangent space of M at x. We denote by $\pi : TM \longrightarrow M$ the canonical projection. A vector field on M is a smooth section of the bundle π, a p-form on M is a smooth section of $\Lambda^p T^* M = \bigcup_{x \in M} \Lambda^p T_x^* M$ where $\Lambda^p T_x^* M$ is the space of p-linear alternating functions

$$\theta_x : \underbrace{T_x M \times \cdots \times T_x M}_{p \text{ times}} \longrightarrow \mathbb{R}.$$

We denote by $\mathfrak{X}_M$ and $\Omega^p(M)$ respectively the space of vector fields and of p-forms. In a coordinate system on an open subset U of M, a p-form θ can be written

$$\theta \equiv \theta|_U = \sum_{i_1 < \cdots < i_p} f_{i_1 < \cdots < i_p} dx_1 \wedge \cdots \wedge dx_p.$$

One defines a $(p+1)$-forms on U by:

$$d_U \theta = \sum_{i_1 < \cdots < i_p} \left(\sum_j \frac{\partial}{\partial x_j} f_{i_1 < \cdots < i_p} dx_{i_j} \right) \wedge dx_{i_1} \wedge \cdots \wedge dx_{i_p}.$$

One shows that the operators d_U fit together into an operator

$$d : \Omega^p(M) \longrightarrow \Omega^{p+1}(M)$$

called the **(de Rham) differential**. It satisfies: $d^2 = d \circ d = 0$.

The space $Z^p(M) = \{\theta \in \Omega^p(M) \text{ s.t } d\theta = 0\}$ is called the space of **cycles** or **closed p-forms** and

$$B^p(M) = \left\{\theta \in \Omega^p(M) \text{ s.t } \exists \, \alpha \in \Omega^{p-1}(M), \theta = d\alpha\right\}$$

is called the "**boundaries**" or "**exact p-forms**".

The equation $d^2 = 0$ implies that $B^p(M) \subset Z^p(M)$. We define:

$$H^p(M, \mathbb{R}) := Z^p(M)/B^p(M)$$

and call $H^p(M, \mathbb{R})$ the pth **de Rham cohomology of** M.

It was proved by de Rham that $H^*(M, \mathbb{R})$ is isomorphic to the singular cohomology of M with real coefficients. Hence $H^*(M, \mathbb{R})$ is an invariant of the manifold M independent of the differentiable structure used to define it.

We also can consider the space $\Omega_c^*(M)$ of forms with compact supports, i.e which vanish identically out some compact subset of M and $Z_c^p(M) = Z^p(M) \cap \Omega_c^p(M)$, $B_c^p(M) = B^p(M) \cap \Omega_c^p(M)$ and define

$$H_c^p(M, \mathbb{R}) := Z_c^p(M)/B_c^p(M)$$

the pth **de-Rham cohomology of** M **with compact supports**.

This cohomology is not even a homotopy invariant (Example: $\mathbb{R}^n$ is homotopy equivalent to a point x, $H_c^n(x) = 0$ but $H_c^n(\mathbb{R}^n) \simeq \mathbb{R}$.)

Here is a list of some basis results:

1. $H^p(M, \mathbb{R}) = \{0\}$ if $p > \dim(M)$ and $p < 0$, $H^0(M, \mathbb{R}) \simeq \underbrace{\mathbb{R} \oplus \cdots \oplus \mathbb{R}}_{k \text{ times}}$;

   where k is the number of connected components of M.

2. Poincaré lemma:

   If $U \subseteq \mathbb{R}^n$ is a star-like open subset of $\mathbb{R}^n$, then $H^p(U, \mathbb{R}) = 0$ for $p > 0$.

3. If M is a n-dimensional compact, oriented manifold without boundary, then $H^n(M, \mathbb{R}) \simeq \mathbb{R}$.

   The isomorphism is induced by the integration

   $$\int \, : \, \Omega^n(M) \longrightarrow \mathbb{R}.$$

4. Stokes theorem:

Let M be an oriented n-dimensional compact manifold with boundary ∂M (which may be empty). Then for any $(n-1)$-form θ one has:

$$\int_M d\theta = \int_{\partial M} \theta.$$

5. Let $S^n = \{x = (x_1, \cdots, x_{n+1}) \in \mathbb{R}^{n+1} \text{ s.t } x_1^2 + \cdots + x_{n+1}^2 = 1\}$ be the unit sphere in $\mathbb{R}^{n+1}$

$$H^p(S^n, \mathbb{R}) = \begin{cases} \mathbb{R} & \text{if } p = 0, n \\ 0 & \text{otherwise.} \end{cases}$$

6. Invariance of the de Rham cohomology by smooth homotopy:

For any $X \in \mathfrak{X}_M$ we define a map $i_X : \Omega^p(M) \longrightarrow \Omega^{p-1}(M)$ called the **interior product by** X:

$$(i_X \theta)\,(\xi_1, \cdots, \xi_{p-1}) = \theta(X, \xi_1, \cdots, \xi_{p-1})$$

for $\theta \in \Omega^p(M)$, $\xi_1, \cdots, \xi_{p-1} \in \mathfrak{X}_M$, and a map

$$\begin{aligned} L_X : \quad \Omega^p(M) \quad &\longrightarrow \quad \Omega^p(M) \\ \theta \quad &\longmapsto \quad L_X \theta = d\,(i_X \theta) + i_X\,(d\theta) \end{aligned} \qquad \text{(A.1.1)}$$

called the **Lie derivative**.

Let X_t be a smooth family of vector fields with compact support on M and let $\varphi_t : M \longrightarrow M$ be the family of diffeomorphisms obtained by integrating the differential equation

$$\frac{d}{dt}\varphi_t(x) = X_t\,(\varphi_t(x))$$

with initial condition $\varphi_0(x) = x$, $\forall\, x$.

If θ_t is a smooth family of differential forms then:

$$\frac{d}{dt}(\varphi_t^* \theta_t) = \varphi_t^* d(i_{X_t}\theta_t) + \varphi_t^* i_{X_t}(d\theta_t) + \varphi_t^*\left(\frac{\partial}{\partial t}\theta_t\right). \quad \text{[Gui-Ste77] (A.1.2)}$$

We rewrite this formula as:

$$\frac{d}{dt}(\varphi_t^* \theta_t) = \varphi_t^*\left(L_{X_t}\theta_t + \frac{\partial}{\partial t}\theta_t\right). \qquad \text{(A.1.3)}$$

In particular (namely if θ does not depend on t and φ_t is symplectic for every t):

$$\varphi_t^*\theta = \theta \Longleftrightarrow L_{X_t}\theta = 0.$$

This formula also expresses the invariance of the de Rham cohomology by the smooth isotopies: integrating the formula (A.1.3), we get:

$$
\begin{aligned}
\varphi_t^*\theta - \theta &= \int_0^t \varphi_s^*(L_{X_s}\theta)ds \qquad\qquad (A.1.4)\\
&= d\left(\int_0^t \varphi_s^*(i_{X_s}\theta)ds\right) + \int_0^t \varphi_s^*\left(i_{X_s}d\theta\right)ds.
\end{aligned}
$$

If $d\theta = 0$, we see that $\varphi_t^*\theta - \theta = d\rho$ where

$$\rho = \int_0^t \varphi_s^*\left(i_{X_s}d\theta\right)ds.$$

This means that the cohomology classes $[\varphi_t^*\theta]$ and $[\theta]$ in $H^p(M,\mathbb{R})$ are equal.

The form $\rho = \rho(\varphi_t)$ above plays an important role in the study of isotopies φ_t preserving a closed p-form θ. In that case

$$d\rho(\varphi_t) = 0$$

and hence defines a cohomology class

$$[\rho(\varphi_t)] \in H^{p-1}(M,\mathbb{R}) \qquad\qquad (A.1.5)$$

which is crucial in the study of symplectic isotopies [Ban78].

A.2 Hodge-de Rham decomposition theorem [War71]

A Riemannian metric g induces an isomorphism

$$\tilde{g} : TM \longrightarrow T^*M$$

like in the symplectic case:

$$\left(\tilde{g}(X)\right)(Y) = g(X,Y).$$

This isomorphism extends to $\Lambda^p TM \longrightarrow \Lambda^p T^* M$. If M is an oriented n-dimensional manifold and v is the corresponding volume form, we define a map

$$\star : \Omega^p(M) \longrightarrow \Omega^{n-p}(M)$$

by $\star\theta = i_{[\tilde{g}^{-1}(\theta)]} v$ where

$$i_{[\tilde{g}^{-1}(\theta)]} v(\xi_1, \cdots, \xi_{n-p}) = v(\tilde{g}^{-1}(\theta), \xi_1, \cdots, \xi_{n-p}).$$

This operator is called the **Hodge-de-Rham star operator**. We define now operators:

$$
\begin{aligned}
\delta &= (-1)^{n(p+1)+1} \star d \star \\
\triangle &= d\delta + \delta d.
\end{aligned}
\qquad\qquad (A.2.1)
$$

The first is called the **co-differential** and the second is called the **laplacian**.

A differential form θ such that $\delta\theta = 0$ is said to be a **co-closed form** and the form such that $\triangle\theta = 0$ is called a **harmonic form**.

The fundamental theorem of Hodge theory is: "*Hodge-de-Rham decomposition theorem*".

Theorem A.1 *(see [War71])*

Let M be a compact oriented manifold. Any p-form θ determines uniquely a harmonic form H_θ and two forms α_1, α_2 with $\delta\alpha_1 = 0 = d\alpha_2$ such that

$$\theta = H_\theta + d\alpha_1 + \delta\alpha_2$$

and if $d\theta = 0$ then $\alpha_2 = 0$.

This decomposition is unique. Moreover, it depends smoothly on θ.

Complete integrability in contact geometry

Augustin Banyaga[1] and Pierre Molino[2]

Abstract

We introduce the notion of completely integrable contact structures as the contact analogs of Duistermaat generalized Lagrangian fibrations in Symplectic Geometry. We construct action-angle coordinates with singularities similar to Eliasson's, define characteristic invariants of completely integrable contact structures like Duistermaat and prove a classification theorem: two completely integrable contact structures with the same characteristic invariants are isomorphic. We study the contact moment map of a torus action preserving a contact form and prove the contact analog of Atiyah-Guillemin-Sternberg convexity theorem, and the contact analog of Delzant realization theorem. The result is applied to the classification of certain K-contact structures.

B.1 Introduction

The conclusions of the Arnold-Liouville theorem and its generalizations lead to view a completely integrable hamiltonian system as a triple $(M^{2n}, \omega, \mathcal{F})$ where M^{2n} is a $2n$-dimensional symplectic manifold with symplectic form ω and a singular foliation $\mathcal{F}$ which is locally defined as the orbits of a local hamiltonian action of the n-torus T^n. The case where this action is global has been studied by Atiyah [3], Guillemin-Sternberg [19], and Delzant [13]. Duistermaat studied the case where the canonical projection on the

---

[1] PennStateUniversity, University Park PA 16802.
[2] Univserty of Montpellier, France.

leave space $\pi : M \longrightarrow W = M/\mathcal{F}$ is a generalized fibration with compact (lagrangian) fibers [16]. The classical Arnold-Liouville theorem describes a neighborhood of a leaf diffeomorphic to T^n and establishes local angle-action coordinates in that neighborhood [1], [23]. Eliasson [17] has studied the situation near a "transversally elliptic" singularity and established local angle-action coordinates with singularities.

Boucetta-Molino [10] and Condevaux-Dazold-Molino [11] have found an integrated formulation and a uniform treatment of all these problems and results, starting with the Arnold-Liouville theorem and Duistermaat fibrations, going through Elliasson's angle-action with elliptic singularities and emerging to the celebrated Atiyah-Guillemin-Sternberg convexity theorem and Delzant's realization theorem.

The next section will give a brief tour of the above point of view of the **complete integrability in Symplectic Geometry** and will develop an analogous theory of **complete integrability in Contact Geometry**.

As a motivation of our study, consider a completely integrable hamiltonian system (M^{2n}, ω, H) with n first integrals $\{H = f_1, f_2, \cdots, f_n\}$ in involution such that $F = \{f_1, \cdots, f_n\} : M \longrightarrow W = H(M) \subset \mathbb{R}^n$ is a proper submersion. Moreover suppose that ω has integral periods and consider a prequantization $\tilde{\pi} : (P, \alpha) \longrightarrow (M, \omega)$ [9], where α is a contact form such that $\tilde{\pi}^*\omega = d\alpha$. Let $H_\alpha = \ker\alpha$ be the contact distribution and consider the submersion $\pi = F \circ \tilde{\pi} : P \longrightarrow W$. We have:

1. $\ker\pi_* \cap H_\alpha$ is a Legendre distribution;

2. the set of all contact gradients X_f, where $f = f'\circ\pi$ with $f' \in C^\infty(W)$, is an abelian Lie algebra $\mathcal{X}$ of infinitesimal automorphisms of H_α and acts transitively on the fibers.

Therefore, the contact analog of a completely integrable hamiltonian system will be a triple (P, H, π) satisfying properties 1 and 2 above. More precisely, we adopt the following:

Definition B.1

A regular completely integrable contact struture on a $(2n+1)$-dimensional manifold P, is a triple (H, π, W) where $H \subset TP$ is a contact structure and $\pi : P \longrightarrow W$ is a proper submersion onto an n-dimensional manifold W such that:

1. $\ker\pi_* \cap H$ is a Legendre distribution;

2. *there exists an abelian Lie algebra $\mathcal{X}$ of infinitesimal automorphisms of the contact structure H, which is transitive on the fibers of π.*

This definition is the Contact Geometry analog of Duistermaat Lagrangian fibrations in Symplectic Geometry.

It is common in the theory of integrable systems to allow singularities of elliptic type in the particular case of compact manifolds. With this in mind, we generalize our definition to allow that type of singularities in contact geometry as well. We are led to the following:

Definition B.2

A completely integrable contact sutructure (with singularities), shortly CIC, is a quadriple $(P, H, \mathcal{X}, \mathcal{A})$ where P is a compact (2n+1)-dimensional manifold with a contact structure $H \subset TP$, $\mathcal{X}$ is an abelian Lie algebra of infinitesimal automorphisms of H and $\mathcal{A}$ is a vector space of first integrals of $\mathcal{X}$ satisfying:

1. $\ker \pi_* \cap H$ *is an isotropic subbundle; here $\pi : P \longrightarrow W$ is the canonical projection on the space of orbits of $\mathcal{X}$;*

2. *the couple $(\mathcal{X}, \mathcal{A})$ is transversally elliptic.*

For the definition of "**transverse ellipticity**" see subsection B.3.2.

The chapter is organized as follows: Section B.2 is a quick review of our vision of the complete integrability in symplectic geometry. In section B.3, we prove the contact analog of Eliasson's theorem: we establish the existence of local angle-action coordinates with singularities for completely integrable contact structures. We treat in detail the regular case (Definition B.1). Note that along the way, we obtain a contact version of the Arnold-Liouville theorem. The key point is that in a neighborhood of a given orbit of $\mathcal{X}$, there exists a T^{n+1} contact action which commutes with elements of $\mathcal{X}$ and has the same orbits. The main results are Theorems **A**, **B** and **C**:

Theorem A

Let $(P, H, \mathcal{X}, \mathcal{A})$ be a CIC-structure on the compact $2n + 1$ dimensional manifold P with an oriented contact structure H. Let α be the contact form obtained in Theorem B.2, which is invariant by $\mathcal{X}$ and by all the T^{n+1}-local actions (see Theorem B.2). Let F_{x_0} be an orbit of $\mathcal{X}$ through x_0 of *dimension* $r + 1$. *Then there exists an open neighborhood* $\mathcal{U}$ *of* F_{x_0} *and a diffeomorphism:*

$$\sigma = (\theta_0, \theta_1, \cdots, \theta_r, q_1, \cdots, q_r, z_1, \cdots, z_{n-r}) :$$

$$\mathcal{U} \longrightarrow T^{r+1} \times \Omega_r \times D_{R_1} \times \cdots \times D_{R_{n-r}}$$

where Ω_r *is an open neighborhooh of* $0 \in \mathbb{R}^r$ *and* D_{R_j} *is an open disk of radius* R_j *in* $\mathbb{C}$ *such that:*

1. $\sigma(x_0) = (\theta_0, \cdots, \theta_r, 0, \cdots, 0)$;

2. $\mathcal{A}$ *is functionally generated by the functions* $(q_1, \cdots, q_r, \rho_1, \cdots, \rho_{n-r})$ *where* $\rho_j = |z_j|$;

3. $(\sigma^{-1})^* \alpha = q_0 d\theta_0 + \sum_{i=1}^{r} q_i d\theta_i + \dfrac{1}{2} \sum_{j=1}^{n-r} \rho_j^2 d\varphi_j$ *where* $\varphi_j = \arg z_j$ *and where* q_0 *is a differentiable function of* q_i's *and of* ρ_j's.

Theorem B

1. *Let* W *be an* n-*dimensional compact manifold with boundary and corners, let* $\mathcal{R}_1$ *be a Legendre lattice on* W *and let* $[\gamma] \in H^2(W, \mathcal{R}_1)$. *Then there exists a completely integrable contact structure* $(P, H, \mathcal{X}, \mathcal{A})$ *with characteristic invariants* $(W, \mathcal{R}_1, [\gamma])$.

2. *If two completely integrable contact structures* $(P, H, \mathcal{X}, \mathcal{A})$ *and* $(P', H', \mathcal{X}', \mathcal{A}')$ *have the same characteristic invariants, then there exists a contact diffeomorphism* $\Phi : P \longrightarrow P'$ *such that* $\pi = \pi' \circ \Phi$ *where* $\pi : P \longrightarrow W$ *and* $\pi' : P' \longrightarrow W$ *are the canonical projections.*

In section B.5, we consider the case where the singular fibration is determined by a global action of T^{n+1}. We prove in that case the contact analog of the celebrated Atiyah-Guillemin-Sternberg convexity theorem and Delzant's realization theorem:

Theorem C

Let (P, H) *be a contact manifold with an effective action of* T^{n+1} *preserving the contact structure. Let* α *be a contact form representing* H, *which is invariant by the* T^{n+1} *action. Let* $\pi : P \longrightarrow W$ *denote the natural projection* *onto the orbit space W; let $J : P \longrightarrow \mathbb{R}^{n+1}$ be the moment map, factoring through $J_W : W \longrightarrow \mathbb{R}^{n+1}$ and $K = J_W(W) \subset \mathbb{R}^{n+1}$ its image.*

1. *Suppose the T^{n+1}-action is regular, then:*

    (a) *W is diffeomorphic to the sphere S^n,*

    (b) *if $n \geqslant 2$ J_W is an embedding which identifies W with the hypersurface K,*

    (c) *if $n \geqslant 3$ then $P \approx T^{n+1} \times S^n$ and the image K of the moment map determines the contact structure H.*

2. *Suppose the T^{n+1}-action is singular and $n \geqslant 2$, then:*

    (a) *the rays in $\mathbb{R}^{n+1}$ from the origin and leaning on K generate a closed convex polytop C,*

    (b) *J_W is an embedding which allows to identify W with the hypersurface with boundary and corners K,*

    (c) *the image K of the moment map determine the contact structure H.*

We finally apply the results to compact K-contact manifolds M^{2n+1} such that the adherence of the flow of the Reeb field is a torus of dimension $n + 1$.

The main ideas of this chapter have been circulated at the "pre-print" level in the Séminaire Gaston Darboux (1991-1992) [6] and got an extensive review in the Mathematical Reviews: MR 94c53029, page 2729.

The following writing has two major improvements over the older preprint: first, we found a more geometric definition of the notion of completely integrable contact structures, expressed purely in terms of contact structures and not in terms of contact forms (Definitions B.1 and B.2). In this formulation the Reeb field plays no significant role; second, we have obtained here a true classification theorem of completely integrable contact structures (Theorem C) while in the previous writing, we had only a sort of "weak classification" of the contact forms up to 1-forms.

The first named author wishes to thank the University of Montpellier and the University of Strasbourg for their hospitality and support during the preparation of this work. Special thanks to Michele Audin and Thomas Delzant for helpful discussions and interest in this work. He also would like to acknowledge partially support by NSF grant DMS 94-03196.

B.2 Complete integrability in symplectic geometry

This chapter is a brief survey of the main results in the geometry of completely integrable hamiltonian system, showing the connections between the classical Arnold-Liouville theorem, Duistermaat's Lagrangian fibrations, the Atiyah-Guillemin-Sternberg theorem of convexity of the moment map of a hamiltonian torus action, and the Delzant realization theorem.

B.2.1 The classical Arnold-Liouville theorem [1], [23]

The classical Arnold-Liouville (A-L) theorem defines angle-action (A-A) coordinates for a hamiltonian system (M^{2n}, ω, H) in a neighborhood $\mathcal{U}$ of a compact connected level set F of n-commuting first integrals $\{g_1 = H, g_2, \cdots, g_n\}$, where $dg_1 \wedge dg_2 \wedge \cdots \wedge dg_n$ is non-zero at a point $x_0 \in F$. The level sets define in $\mathcal{U}$ a compact lagrangian foliation, angle action correspond to a free hamiltonian action of the torus T^n, while action variables are the components of the moment map of this action.

The existence problem for global A-A coordinates on a symplectic manifold endowed with a proper lagrangian fibration, studied by Duistermaat [16], led to the definition of characteristic invariants on the quotient: an integral affine structure determined by local action variables, and a Chern class of the fibration.

On the other hands, convexity properties for hamiltonian action of tori on closed symplectic manifolds where obtained by Atiyah [3] and Guillemin-Sternberg [19], using an adapted version of Bott-Morse theory. For an elegant exposition, see M. Audin [4]. If the dimension of the torus is half the dimension of the manifold, the convexity theorem is completed by Delzant's important result [13]: the convex polytop, which is the image of the moment map, completely determines - up to equivalence - the symplectic manifold and the toral action.

The neighborhood of a degenerate orbit of a hamiltonian action of T^n on (M^{2n}, ω) is a natural model of the so-called **"elliptic singularities"** for completely integrable systems; from this point of view, the singular version of A-L theorem is the following result due to Eliasson [17]: let $g_1, \cdots, g_n$ be commuting first integrals of $(M^{2n}, \omega, H = g_1)$ and x_0 a point in M with $dg_1(x_0) \wedge \cdots \wedge dg_k(x_0) \neq 0$ and $dg_{k+1}(x_0) = \cdots = dg_n(x_0) = 0$. Assume that F is a compact connected level set of $(g_1, \cdots, g_n)$ with $x_0 \in F$ define the transverse space $N_T(x_0)$ at x_0 as:

$$N_T(x_0) = \bigcap_{i=1}^{k} \ker dg_i(x_0)/\mathcal{T}(x_0)$$

where $\mathcal{T}(x_0)$ is the subspace of $T_{x_0}(M)$ spanned by $X_{g_j}(x_0), j = k + 1, \cdots, n$. Here we denoted by X_f the hamiltonian vector field of a function f on (M, ω). The linear parts of $X_{g_{k+1}}, \cdots, X_{g_n}$ at x_0 induce linear vector fields $X'_{k+1}, \cdots, X'_n$ on $N_T(x_0)$. If these linear vector fields generate an effective action of T^{n-k} then there exists in a neighborhood $\mathcal{U}$ of F a hamiltonian action of T^n whose orbits are the level sets of $(g_1, \cdots, g_n)$. Moreover there is on $\mathcal{U}$, angle-action coordinates "with singularities". The A-L theorem correspond to the case $k = n$.

B.2.2   A unified theory including both global and singular properties

The first step in that direction was done in [11], where an alternative proof of the Atiyah-Guillemin-Sternberg convexity theorem was given, based on the following idea: if (M^{2n}, ω) is endowed with a hamiltonian action of T^k, one considers the space of connected components of the - regular or singular - fibers of the moment map. This quotient space inherits a natural structure of flat riemannian manifold with boundary and corners, moreover, local convexity properties of the moment map obtained from standard models in the neighborhood of singular orbits, imply a geodesic convexity of this singular quotient manifold; hence, the canonical projection defined by the moment map appears to be an inclusion of the quotient space as a convex polytop in $\mathbb{R}^k$.

If $k = n$, this point of view leads to a general notion of lagrangian compact foliation with elliptic singularities, that is to say, an intrinsic notion of completely integrable hamiltonian system with elliptic singularities, where, instead of n particular first integrals, one considers a commuting Lie algebra of first integrals. This type of structure has been studied in [10], the main result being, a classification, via characteristic invariants of the quotient space, which is a locally convex integral affine manifold integral boundaries and corners. This result generalizes Duistermaat's as well as Atiyah-Guillemin-Sternberg's and Delzant's theorems.

Finally, a more general notion of "elliptic singularity" was studied in [15], where a new proof of Eliasson's result is given by using a blowing up procedure. The key point is the following: let M be an arbitrary manifold and $\mathcal{X}$ be an abelian Lie algebra of vector fields on M, $\mathcal{A}$ a ring of first integrals of $\mathcal{X}$, F a compact orbit of $\mathcal{X}$, x_0 a point of F. The ellipticity at x_0 of the pair $(\mathcal{X}, \mathcal{A})$ is defined exactly as in the hamiltonian case, where $\mathcal{A}$ is generated by the hamiltonians $(g_1, \cdots, g_n)$ and $\mathcal{X}$ is generated by the associated symplectic gradients. If this condition is satisfied, the conclusion is that there exists in a neighborhood $\mathcal{U}$ of F, an action of the torus T^n which commutes with $\mathcal{X}$ and has the same orbits (Dufour-Molino compactification theorem [15]).

From this point of view, a completely integrable hamiltonian system with elliptic singularities on (M^{2n}, ω) will be defined as a pair $(\mathcal{X}, \mathcal{A})$, where $\mathcal{X}$ is an abelian Lie algebra of local hamiltonian vector fields, and $\mathcal{A}$ is a Poisson-commuting ring of functions such that the pair $(\mathcal{X}, \mathcal{A})$ satisfy the ellipticity at each point.

Starting from this point of view, it is easy to understand how to define a natural notion of completely integrable contact structure with elliptic singularities. The purpose of this work is precisely to study the contact version of the previous results.

B.3   Contact angle-action coordinates

B.3.1   Contact geometry preliminaries [7], [8], [23]

A **contact form** on a $(2n+1)$-dimensional smooth manifold P is a 1-form α such that $\alpha \wedge (d\alpha)^n$ is everywhere non-zero. A **contact structure** on a smooth manifold P is a hyperplane field $H \subset TP$ of the tangent bundle such that each point $x \in P$ has an open neighborhood U such that there exists a contact form α_U defined on U the kernel of which is the restriction H_U of H over U. The couple (P, H) is called a contact manifold.

If the contact manifold is oriented, then there exists a global contact form (defined on the entire manifold) α such that $\ker \alpha = H$. We say that H is defined by α and write $H = [\alpha]$. Two contact forms α and α' define the same contact structure H if and only if $\alpha = \lambda \alpha'$ for some nowhere vanishing function λ.

In this paper we will consider only oriented contact structures.

If $H = [\alpha]$ for some contact form α, then by definition, the restriction of $d\alpha$ to H is a symplectic structure, i.e. a non-degenerate bilinear two form. In fact $d\alpha$ is a 2-form of rank $2n$. There is a unique global section Z of its kernel normalized by the condition that $\alpha(Z) = 1$. The vector field Z is called the **Reeb field** of α. It is uniquely characterized by the equations:

$$\begin{cases} i_Z\alpha & = & 1 \\ i_Z d\alpha & = & 0 \end{cases}$$

where i_Z stands for the interior product by Z. Notice that the Reeb field is not an invariant of the contact structure. If $\alpha' = \lambda\alpha$, the Reeb field of α is not nicely related to the Reeb field of α' [7], [23].

To each vector field X, we can assign a section X_H of H, (those sections are called "horizontal vector fields", or the horizontal part of X) defined by:

$$X_H = X - (i_X\alpha)Z.$$

And for each horizontal vector field Y, there exists a uniquely defined 1-form β_Y such that

$$\beta_Y = i_Y d\alpha$$

such forms are called semi-basic 1-forms (those whose interior product with Z is identically zero). There is a $1 - 1$ correspondance between semi-basic 1-forms and horizontal vector fields [7], [23].

Let $C^\infty(P)$ denote the set of all smooth functions on P. For each $f \in C^\infty(P)$, $(i_Z df)\alpha - df$ is a semi-basic 1-form and hence gives rise to a horizontal vector field H_f such that

$$i_{H_f} d\alpha = (i_Z df)\alpha - df.$$

A **contact diffeomorphism** between two contact manifolds (P, H) and (P', H') is a diffeomorphism $h : P \longrightarrow P'$ such that $h_*H = H'$. If $H = [\alpha]$ and $H' = [\alpha']$, then $h^*\alpha' = \lambda\alpha$ for some nowhere zero function λ. An **infinitesimal automorphism** of a contact structure (P, H) is a vector field X, called a "contact vector field" such that its local 1-parameter group is made of contact diffeomorphisms. If $H = [\alpha]$, a contact vector field X satisfies:

$$L_X\alpha = \mu\alpha$$

for some function μ and here L_X stands for the Lie derivative in the direction X. Let $\mathcal{X}(H)$ be the set of all contact vector fields. Suppose that

$H = [\alpha]$. The mapping:

$$a: \quad \begin{array}{ccc} \mathcal{X}(H) & \longrightarrow & C^\infty(P) \\ X & \longmapsto & i_X \alpha \end{array}$$

is an isomorphism whose inverse is: $f \longmapsto X_f =: H_f + fZ$. It is easy to verify that:

$$L_{X_f} \alpha = (i_Z df) \alpha.$$

The vector field X_f is a contact vector field with $i_{X_f} \alpha = f$.

Notice that if $Z \cdot f = i_Z df = 0$, then $L_{X_f} \alpha = 0$. Such a vector field is said to be a strictly contact vector field. Its local 1-parameter group preserves the contact form. The set of strictly contact vector fields is a Lie subalgebra of $\mathcal{X}(H)$ isomorphic to the set of "basic functions": those functions which are invariant under the flow of the Reeb field, such that $Z \cdot f = 0$.

Let us fix a contact form α on a contact manifold (P, H) such that $H = [\alpha]$, and let Z be the Reeb field of α.

There are (infinitely many) riemannian metrics g and 1-1 tensor fields J such that: $JZ = 0$ and $J^2 = -I$ on H and satisfying the following conditions:

$$\begin{cases} g(X, Y) & = & g(JX, JY) + \alpha(X)\alpha(Y) \\ d\alpha(X, Y) & = & g(X, JY) \end{cases}$$

for all vector fields X, Y. This means that "transversally to the Reeb field", i. e. on H, the almost complex structure J is tamed by $d\alpha$.

As a consequence of the formulas above, we get:

$$\alpha(X) = g(X, Z).$$

Such metrics are called **contact metrics** [7], [8]. Using a contact metric, the contact distribution H appears as the orthogonal complement to the 1-dimensional distribution spanned by the Reeb field.

We say that the contact form α is a K-**contact form** if the Reeb field is Killing with respect to some contact metric g. Such contact forms appear naturally in many settings. For instance Brieskorn manifolds carry K-contact structures.

K-contact structures have the property that the Reeb field is almost periodic, i.e. the closure of its flow is a torus acting on the contact manifold preserving the contact form. We will analyze this in details in section 5.

Finally let us recall the definition of **Legendre distribution**. A **Legendre submanifold** of a $(2n + 1)$ dimensional contacy manifold (P, H) is an n-dimensional submanifold L such that $T_x L \subset H_x$ for all $x \in L$. If $H = [\alpha]$ for some contact form α and if $l : L \longrightarrow P$ is the embedding of L into P, then $l^*\alpha = 0$. Hence also $l^*d\alpha = d(l^*\alpha) = 0$. Therefore $l_* T L$ is a lagrangian subbundle of H. Hence we call a **Legendre distribution** of (P, H) any Lagrangian subbundle of H. This notion is well defined: namely, it is independent of the choice of the contact form α such that $H = [\alpha]$, since the conformal class of the transverse symplectic structure $d\alpha$ is a contact invariant.

B.3.2   The regular case

After these brief contact preliminaries, let us start the core of this paper.

Recall (Definition 1 in the introduction) that a **regular completely integrable contact structure** on a smooth $(2n+1)$-dimensional manifold P consists of:

1. a contact structure $H \subset TP$.

2. a proper submersion $\pi : P \longrightarrow W$ onto an n-dimensional manifold W, with connected fibers and such that $V = \ker \pi_* \cap H$ is a Legendre distribution.

3. an abelian Lie algebra $\mathcal{X}$ of infinitesimal automorphisms of H, which has the fibers of π as orbits.

In particular if $x \in P$ and if $\mathcal{X}_x$ denotes the set $\{X_x = X(x), X \in \mathcal{X}\}$, then $\ker(T_x \pi) = \mathcal{X}_x$.

Sometime it will be more convenient to consider the "completion" $\widetilde{\mathcal{X}}$ of $\mathcal{X}$ defined as the set of all infinitesimal automorphisms of H, belonging to $\ker \pi_*$ and commuting with all elements of $\mathcal{X}$. The triple $(P, H, \mathcal{X})$ or $(P, H, \widetilde{\mathcal{X}})$ are called a **regular completely integrable contact manifold**. We abbreviate this notion by the RCIC-manifold.

Proposition B.1

   Let $(P, H, \mathcal{X})$ be an RCIC-manifold. Then $\ker \pi_*$ is everywhere transverse to H, i.e. for each $x \in P$, then $T_x P = \ker(T_x \pi) + H_x$.

Proof

The dimension of $V = \ker \pi_* \cap H$ is n, (since V is a Legendre distribution and the dimension of P is $2n+1$). Since $\dim H = 2n$ and $\dim \ker \pi_* = n+1$, it follows that $\dim(\ker \pi_* + H) = (n+1) + (2n) - (n) = 2n + 1$. $\qquad\square$

Proposition B.2

1. *Each point of P has an open neighborhood on which the contact structure H can be represented by a (local) contact form which is invariant by all elements of $\mathcal{X}$.*

2. *If the contact structure H is oriented, then there exists a global contact form, representing H and invariant by all elements of $\mathcal{X}$.*

Proof

1. According to proposition B.1, for each point x_0 of P, there exists $X \in \mathcal{X}$ such that X_{x_0} is transverse to H_{x_0}. This vector field X stays transverse to H on a whole neighborhood U of x_0. Pick any contact form α_0 defined on U (or an eventually smaller open neighborhood of x_0) representing H, i.e. $H_{|U} = \ker \alpha_0$. Since X is transverse to H on U, $\alpha_0(X) \neq 0$ on U. Then $\alpha = \dfrac{\alpha_0}{\alpha_0(X)}$ is the local contact form required in (1). Indeed, let $Y \in \mathcal{X}$, then $[X, Y] = 0$ since $\mathcal{X}$ is abelian. On the other hands, $L_Y \alpha = \phi \alpha$ for some function ϕ since X is an infinitesimal automorphism of H. Therefore:

$$0 = i_{[X,Y]}\alpha = L_Y i_X \alpha - i_X L_Y \alpha = -\phi$$

since $i(X)\alpha = 1$. Here $i(.)$ denotes the interior product and L_Y stands for the Lie derivative in direction Y. Hence the contact form α is invariant by $\mathcal{X}$.

Notice also that $i_X d\alpha = L_X \alpha = 0$. Hence X is the Reeb field of α (on U). Let us denote by Z_U its restriction to U.

In fact α and Z_U are defined on $\hat{U} = \pi^{-1}\big(\pi(U)\big)$ since α is invariant by $\mathcal{X}$ and the orbits of $\mathcal{X}$ are the fibers of the submersion $\pi : P \longrightarrow W$.

2. Let $\{V_i\}$ be an open cover of W, chosen so fine that on $U_i = \pi^{-1}(V_i)$, we can define, using (i) contact forms α_i which are invariant by $\mathcal{X}$. Let Z_i be the corresponding Reeb fields and let $\{\lambda_i\}$ be a partition of

unity subordinate to $\{V_i\}$. The assumption that the contact structure is oriented means that on $U_i \cap U_j$, $\alpha_i = f_{ij}\alpha_j$ with $f_{ij} \geqslant 0$. Under that assumption, the 1-form:

$$\alpha = \sum_i (\lambda_i \circ \pi)\alpha_i$$

is a contact form.

Indeed, on an open set, say U_1, $\alpha = \lambda\alpha_1$, where $\lambda = \sum_i (\lambda_i \circ \pi)f_{1i} \geqslant 0$ on U_1, summing over all indices i such that U_i meets U_1.

For $X \in \mathcal{X}$, we have: $L_X\alpha = \sum (X \cdot (\lambda_i \circ \pi))\alpha_i + (\lambda_i \circ \pi)L_X\alpha_i = 0$ since X preserves all α_i and is π-vertical. $\qquad\square$

Remark B.1

1. *Let Z be the Reeb field of α. For any $Y \in \mathcal{X}$, then $[Y, Z] = 0$. Indeed,*

$$i_{[Y,Z]}\alpha = i_Z L_Y \alpha - i_Y L_Z \alpha = 0.$$

   We conclude that $[Y, Z] = 0$ since $a([Y, Z]) = 0$ and a is an isomorphism (see 3.1).

2. *The abelian Lie algebra $\hat{\mathcal{X}}$ generated by Z and $\mathcal{X}$ (the completion of $\mathcal{X}$) acts on P, preserving the local contact forms. Since $\mathcal{X}$ has already orbits of maximum dimension $(n+1)$, the vector field Z is everywhere a linear combination of vector fields of $\mathcal{X}$. Hence Z is π-vertical.*

3. *The π-verticality of Z implies that $i_Z d(f \circ \pi) = 0$ for all $f \in C^\infty(W)$. Therefore, we can define a unique section H_f of H by:*

$$i_{H_f}d\alpha = -d(f \circ \pi).$$

   The vector field X_f on P defined by:

$$X_f = fZ + H_f$$

   satisfies $L_{X_f}\alpha = 0$. The vector field X_f is called the contact gradient of f.

4. *For any $Y \in \mathcal{X}$, the function $i_Y \alpha$ is π-basic. Indeed, for any $X \in \mathcal{X}$,*

$$X \cdot \alpha(Y) = L_X i_Y \alpha = i_Y L_X \alpha = 0$$

since $[X, Y] = 0$ and $L_X \alpha = 0$. Let $f_Y \in C^\infty(W)$ such that $f_Y \circ \pi = i_Y \alpha$, then clearly

$$Y = X_{f_Y}$$

which means that all elements of $\mathcal{X}$ are contact gradients of some smooth functions on W.

Lemma B.1

Given $g \in C^\infty(W)$ and $x_0 \in W$, there exist an open neighborhood V of x_0 in W, $X_1, \cdots, X_n \in \mathcal{X}$ and $(n+1)$ functions $f_1, \cdots, f_n$ on V such that:

$$X_g|_{\pi^{-1}(V)} = (f_0 \circ \pi)Z + \sum_{i=1}^{n}(f_i \circ \pi)X_i.$$

Proof

Choose $X_1, \cdots, X_n \in \mathcal{X}$ such that $Z(y), X_1(y), \cdots, X_n(y)$ are linearly independent for $y \in \pi^{-1}(x_0)$. Let g_k be the functions such that $g_k \circ \pi = i(X_i)\alpha$; there exists an open neighborhood $V \subset W$ of x_0 where $dg_1, \cdots, dg_n$ are functionally independent.

Now given $g \in C^\infty(W)$, there are functions $f_1, \cdots, f_n \in C^\infty(W)$ such that

$$dg = \sum_{i=1}^{n} f_i dg_i.$$

Define

$$f_0 = g - (f_1 g_1 + f_2 g_2 + \cdots + f_n g_n).$$

And consider

$$X = f_0 Z + \sum_{i=1}^{n} f_i X_i.$$

We have:

$$\begin{cases} i_X \alpha &= g \\ i_X d\alpha &= \displaystyle\sum_{i=1}^{n} f_i(-dg_i) = -dg. \end{cases}$$

Hence $X = X_g$.   □

The remarks above and Lemma B.1 imply the following:

Proposition B.3

   The completion $\widehat{\mathcal{X}}$ of $\mathcal{X}$ locally consists of contact gradients of local basic functions.

The main result of this section is the following:

Theorem B.1

   Let $y_0 \in W$ and $x_0 \in \pi^{-1}(y_0)$. There exist an open neighborhood U_W of $y_0 \in W$ and a free action of T^{n+1} on $U = \pi^{-1}(U_W)$, the orbits of which are the fibers of π, which commutes with $\mathcal{X}$ and which preserve the contact form $\alpha_U = \alpha|_U$. Moreover there exist coordinates $(q_1, \cdots, q_n, \theta_0, \cdots, \theta_n)$ identifying U with $U_W \times T^{n+1}$ and in which the contact form $\alpha|_U$ assumes the following expression:

$$\alpha|_U = q_0 d\theta_0 + q_1 d\theta_1 + \cdots + q_n d\theta_n$$

where q_0 is a smooth function of $q_1, \cdots, q_n$.

Proof

   The existence of a local torus action which has the fibers as orbits and commuting with $\mathcal{X}$ is standard. The point is that on each fiber $\pi^{-1}(y_0)$, $\mathcal{X}$ induces a transitive action of $\mathbb{R}^{n+1}$ and the isotropy subgroup I_x of each point X in the fiber depends differentiably on $\pi(x)$ and it is a lattice in $\mathbb{R}^{n+1}$. The T^{n+1} action then follows by quotienting $\mathbb{R}^{n+1}$ by the lattice. See [23], section 16.10.

   At this stage, we choose coordinates $(x_1, \cdots, x_n, \tilde{\theta}_0, \cdots, \tilde{\theta}_n)$ on U identifying U with $U_W \times T^{n+1}$. In those coordinates we have:

$$\frac{\partial}{\partial \tilde{\theta}_k} = \sum_j f_{kj} X_j$$

where f_{kj} are basic functions.

   Since $\alpha(X_k)$ is π-basic, so is $i_{X_k} d\alpha = -d(\alpha(X_k))$.

   Therefore: $i_{\frac{\partial}{\partial \tilde{\theta}_k}} d\alpha = \sum_j f_{kj} i_{X_j} d\alpha$ is a basic 1-form. Hence, evaluating it to the "vertica" vector field $\frac{\partial}{\partial \tilde{\theta}_l}$, we get

$$d\alpha \left(\frac{\partial}{\partial \tilde{\theta}_k}, \frac{\partial}{\partial \tilde{\theta}_l} \right) = 0.$$

This fact implies that $d\alpha$ has the following expression:

$$d\alpha = \sum_{i,\nu} A_{i\nu}(x)d\tilde{\theta}_i \wedge dx_\nu + \pi^*\beta$$

where $A_{i\nu}$ and β are basic. The fact that $d(d\alpha) = 0$ implies that

$$d\left(\sum_\nu A_{i\nu}dx_\nu\right) = 0.$$

We can then change variables in U_W so that

$$d\alpha = \sum_i d\tilde{\theta}_i \wedge d\tilde{q}_i + \pi^*\eta.$$

Setting $\eta = \sum_i f_i d\tilde{q}_i$ and changing the variables $\tilde{\theta}_i$ to $\hat{\theta}_i = \tilde{\theta}_i + f_i$ gives $d\alpha$ the expression:

$$d\alpha = \sum_{i=1}^n d\hat{\theta}_i \wedge d\tilde{q}_i.$$

Therefore:

$$\alpha = \alpha_0 + dh$$

where

$$\alpha_0 = -\sum_{i+1}^n \tilde{q}_i d\hat{\theta}_i$$

and h is a π-basic function.

Let Z be the Reeb field of α, we have

$$\begin{cases} i_Z\alpha &= i_Z\alpha_0 = 1 \\ i_Z d\alpha &= i_Z d\alpha_0 = 0. \end{cases}$$

Hence $L_Z\alpha = L_Z\alpha_0 = 0$. Consider now the vector field $Y = -hZ$, where h is the π-basic function above. The vector field Y is complete since Z was and its flow φ_t commutes with the torus action.

Let $\alpha_t = \alpha_0 + tdh$. We have:

$$L_Y\alpha_t = L_Y\alpha_0 + tY \cdot h = L_Y\alpha_0 = -dh = -\frac{\partial}{\partial t}\alpha_t.$$

Therefore:

$$\frac{d}{dt}(\varphi_t^*\alpha_t) = \varphi_t^*\left(L_Y\alpha_t + \frac{\partial}{\partial t}\alpha_t\right) = 0$$

which implies that $\varphi_1^* \alpha_1 = \varphi_1^* \alpha = \alpha_0$. This last change of coordinates takes α to the desired form α_0. On this expression, it is clear that $L_{\frac{\partial}{\partial \theta_k}} \alpha_0 = 0$.

$\square$

The goal of the next section is to show that the conclusion of theoren 1 hold in general for completely integrable contact structures with singularities. Recall that the main points are the existence of a global contact form α and local torus actions having the fibers of π as orbits, commute with $\mathcal{X}$ and preserve the contact form α.

B.3.3 The singular case

The notion of CIC-structures

The definition of completely integrable contact structures (CIC-structures) was given in the introduction. We recall it here.

Definition B.3

A completely integrable contact structure on a $(2n+1)$-dimensional manifold P consists of:

1. *a contact structure $H \subset TP$ on P,*

2. *an abelian Lie algebra $\mathcal{X}$ of infinitesimal automorphisms of H,*

3. *a vector space $\mathcal{A}$ of first integrals of $\mathcal{X}$ subject to the following conditions:*

    (a) $V = \ker \pi_* \cap H$ is an isotropic subbundle of H. Here $\pi : P \longrightarrow W$ is the canonical projection onto the orbit space W of $\mathcal{X}$.

    (b) The couple $(\mathcal{X}, \mathcal{A})$ is transversally elliptic at every point.

The notion of "transverse ellipticity" will be treated in the next section. Let us now state the analogous of Proposition B.2.

Theorem B.2

Let $(P, H, \mathcal{X}, \mathcal{A})$ be a CIC-structure on the compact manifold P with an oriented contact structure H. Then each point of P has an open neighborhood U and an action of T^{n+1} on U, having the same orbits as $\mathcal{X}$, and commuting with each element of $\mathcal{X}$.

Furthermore, the contact structure H can be represented by a global contact form α which is invariant by $\mathcal{X}$ and by all the T^{n+1} local actions.

The notion of transverse ellipticity [15]

Let $\mathcal{X}$ be a commutative Lie algebra of vector fields on a smooth manifold P and let $\mathcal{A}$ be a vector space of first integrals of $\mathcal{X}$, i.e. functions $f \in \mathcal{A}$ such that $X \cdot f = 0$, $\forall X \in \mathcal{X}$. This also means that if we denote by $\mathcal{X}_{x_0}$, $x_0 \in P$ the subspace of $T_{x_0}P$ formed by $\{X_{x_0}\}$, $X \in \mathcal{X}$, then $\mathcal{X}_{x_0}$ is a subspace of $K_{x_0} = \bigcap_{f \in \mathcal{A}} \ker(d_{x_0}f)$. The quotient space:

$$N_{x_0} = K_{x_0}/\mathcal{X}_{x_0}$$

is called the **transverse space**. Let J_{x_0} be the subspace of $\mathcal{X}$ formed by those vector fields vanishing at x_0, i.e. the *"isotropy"* of the point x_0. The linear part X'_{x_0} of $X \in J_{x_0}$ acts on $T_{x_0}P$ preserving K_{x_0} and $\mathcal{X}_{x_0}$, hence induces a linear transformation $X_T'^{x_0}$ of the transverse space N_{x_0}. We denote by $\mathcal{X}_T'^{x_0}$ the abelian Lie algebra formed by all the linear transformations $X_T'^{x_0}$; $X \in J_{x_0}$. Likewise let $\mathcal{A}_{x_0}$ be the subspace of $\mathcal{A}$ of functions which are critical at x_0. The kernel of the Hessian of $f \in \mathcal{A}_{x_0}$ contains $\mathcal{X}_{x_0}$. Therefore, each Hessian of $f \in \mathcal{A}_{x_0}$ induces a quadratic form $f_T'^{x_0}$ on N_{x_0}. We denote by $\mathcal{A}_T'^{x_0}$ the set of all those quadratic forms.

Definition B.4

   The couple $(\mathcal{X}, \mathcal{A})$ is said to be **transversally elliptic** *(T.E) at x_0 if N_{x_0} has a symplectic vector space structure such that $\mathcal{X}_T'^{x_0}$ is the Lie algebra of a maximum torus in the symplectic group $Sp(N_{x_0})$ and $\mathcal{A}_T'^{x_0}$ is the Poisson algebra of corresponding hamiltonians.*

   The symplectic vector space on N_{x_0} is obtained by a *"reduction"* procedure. Let $2s$ be the dimension of N_{x_0}, the transversality condition can also be rephrased as follows:

Proposition B.4

   There are linear coordinates $(x_1, \cdots, x_s, y_1, \cdots, y_s)$ on N_{x_0} with respect to which a basis of $\mathcal{X}_T'^{x_0}$ is given by the following infinitesimal rotations:

$$\left\{ y_j \frac{\partial}{\partial x_i} - x_j \frac{\partial}{\partial y_i} \quad i, j = 1, \cdots, s \right\}$$

and a basis of $\mathcal{A}_T'^{x_0}$ is given by the quadratic forms:

$$\{x_j^2 + y_j^2 ; \, j = 1, \cdots, s\}.$$

The interest of this notion is due to the following result:

Theorem B.3 *(Dufour-Molino compactification theorem [15])*
Let $(\mathcal{X}, \mathcal{A})$ as above and $x_0 \in P$. Assume the orbit $\mathcal{O}_{x_0}$ of $\mathcal{X}$ through x_0 is compact and that $(\mathcal{X}, \mathcal{A})$ is transversally elliptic at x_0. Then there exists an open neighborhood U of x_0 saturated of $\mathcal{X}$-orbits and a T^n action on U which commutes with each element of $\mathcal{X}$ and such that the T^n-orbits coincide with the $\mathcal{X}$-orbits.

Our Theorem B.2 will be a consequence of the compactification theorem and of the next two propositions:

Proposition B.5 *give up*
Suppose the manifold P is compact and $(\mathcal{X}, \mathcal{A})$ is transversally elliptic at x_0, then $\mathcal{O}_{x_0}$ is compact (hence a torus T^k).

Proof
Let us call a "level" of $\mathcal{A}$ and denote it by $L(\mathcal{A})$ any subset of P on which all the functions $f \in \mathcal{A}$ keep the same value (and which are maximal for this property). The levels $L(\mathcal{A})$ are compact subsets of P since P is compact. Here $\mathcal{A}$ is the given vector space of first integrals of $\mathcal{X}$. Clearly, orbits of $\mathcal{X}$ are subsets of $L(\mathcal{A})$.

Let x_0 be a point in P such that the orbit $\mathcal{O}_{x_0}$ has dimension $r + 1$. The hypothesis of transverse ellipticity at x_0 implies that there exist an open neighborhood U of x_0 and local coordinates centered at x_0:

$$u_0, u_1, \cdots, u_r, v_1, \cdots, v_r, (x_i, y_i)_{i=1 \cdots n-r}$$

such that $U \cap \mathcal{O}_{x_0}$ is given by the following equations:

$$v_j = x_i = y_i = 0, j = 1, \cdots, r; \quad i = 1, \cdots, n-r.$$

See [14], and [15]. This implies that the orbit is closed in the the level. Hence it is compact. □

Proposition B.6
Let (P, H) be a contact manifold with an abelian Lie algebra $\mathcal{X}$ of infinitesimal automorphisms of H, with local T^{n+1} actions, preserving H, commuting with $\mathcal{X}$ such that the T^{n+1}-orbits coincide with the $\mathcal{X}$-orbits. If the map $x \mapsto T\mathcal{O}_x \cap H_x$ is an isotropic subbundle, then $T\mathcal{O}_x$ is everywhere transverse to H_x.

Proof

Let x_0 be a point so that the $\mathcal{X}$-orbit $(T^{n+1}$-orbit) has dimension $n+1-k$, i.e. the isotropy I_{x_0} is k-dimensional. Suppose that $\mathcal{O}_{x_0}$ is tangent to H_{x_0}. The tangent space $E_{x_0} \subset H_{x_0}$ to $\mathcal{O}_{x_0}$ is an isotropic subbundle of H_{x_0}, by assumption. The linearization I'_{x_0} of I_{x_0} acts on $T_{x_0}P$ inducing an automorphisms of H_{x_0}, and also of $H_{x_0}^{\perp}$. Since the later is one dimensional, we conclude that I'_{x_0} acts faithfully on H_{x_0}. Recall that dim $(E_{x_0}) = n+1-k$ so that dim $E_{x_0}^{\perp} = 2n - (n+1-k) = n+k+1$ and the reduced space $\overline{H}_{x_0} = E_{x_0}^{\perp}/E_{x_0}$ is a symplectic vector space of dimension $2k-2$, on which the torus $T^k = I'_{x_0}$ acts symplectically. This is impossible since the maximum dimension of a torus acting faithfully symplectically on a $(2k-2)$ dimensional symplectic manifold is $k-1$. This contradiction establishes that the orbit $\mathcal{O}_{x_0}$ is transverse to H_{x_0}. $\square$

Proof of Theorem B.2

According to Proposition B.6, we can apply the compactification theorem. Near a regular point, i.e. a point x_0 such that $\mathcal{O}_{x_0}$ has the maximal dimension $n+1$, the compactification is nothing but what we did in Theorem B.1. Let $U \subset P$ be the open set in the compactification theorem and $\pi : U \longrightarrow W$, the projection onto the orbit space W. We found in theorem 1 coordinates $(q_1, \cdots, q_n, \theta_0, \theta_1, \cdots, \theta_n)$ on $W \times T^{n+1}$ identifying U with $W \times T^{n+1}$ with the natural action of T^{n+1} on $W \times T^{n+1}$, so that the contact structure H can be represented on U by the contact form:

$$\alpha = q_0 d\theta_0 + q_1 d\theta_1 + \cdots + q_n d\theta_n$$

which is obviously invariant by all the generators $\frac{\partial}{\partial \theta_k}$ of the torus action. Hence this (modified) compactification respects the contact structure H in the open set U, a neighborhood of a regular point x_0. Since the set of regular points is everywhere dense in P, the local actions of T^{n+1} respect H everywhere by continuity.

The last thing to be proved is the existence of a global contact form representing H which is invariant by all the local T^{n+1}-actions. As in the regular case, we choose an open cover $\{W_i\}$ of the orbit space W by open sets W_i small enough so that there exists for each i a vector field $X_i \in T\mathcal{O}$ transverse to H on $\pi^{-1}(W_i)$. Exactly like in the proof of Theorem B.1, there are contact forms α_i on $\pi^{-1}(W_i)$ such that $\alpha_i(X_i) = 1$ and invariant by T^{n+1}. We saw that if $\{\lambda_i\}$ is a partition of unity subordinate to $\{W_i\}$, then $\alpha = \sum (\lambda_i \circ \pi)\alpha_i$ is an invariant contact form. $\square$

The main result of this chapter is the following generalization of Theorem 1:

Theorem A

Let $(P, H, \mathcal{X}, \mathcal{A})$ be a CIC-structure on the compact $2n + 1$ dimensional manifold P with an oriented contact structure H. Let α be the contact form obtained in Theorem B.2, which is invariant by $\mathcal{X}$ and by all the T^{n+1}-local actions (see Theorem B.2). Let F_{x_0} be an orbit of $\mathcal{X}$ through x_0 of dimension $r + 1$. Then there exists an open neighborhood $\mathcal{U}$ of F_{x_0} and a diffeomorphism:

$$\sigma = (\theta_0, \theta_1, \cdots, \theta_r, q_1, \cdots, q_r, z_1, \cdots, z_{n-r}) :$$

$$\mathcal{U} \longrightarrow T^{r+1} \times \Omega_r \times D_{R_1} \times \cdots \times D_{R_{n-r}}$$

where Ω_r is an open neighborhood of $0 \in \mathbb{R}^r$ and D_{R_j} is an open disk of radius R_j in $\mathbf{C}$, such that

1. *$\sigma(x_0) = (\theta_0, \cdots, \theta_r, 0, \cdots, 0)$,*

2. *$\mathcal{A}$ is functionally generated by the functions $(q_1, \cdots, q_r, \rho_1, \cdots, \rho_{n-r})$ where $\rho_j = |z_j|$,*

3. *$(\sigma^{-1})^* \alpha = q_0 d\theta_0 + \displaystyle\sum_{i=1}^{r} q_i d\theta_i + \frac{1}{2} \sum_{j=1}^{n-r} \rho_j^2 d\varphi_j$*

   where $\varphi_j = \arg z_j$ and where q_0 is a differentiable function of q_i's and of ρ_j's.

By analogy with the symplectic case [10], we will say that $(\theta, q, \varphi, \frac{1}{2}\rho^2)$ are contact singular angle-action coordinates (CSAAC) for the CIC-structure $(P, H, \mathcal{X}, \mathcal{A})$, where (θ, φ) are the "angles" and $(q, \frac{1}{2}\rho^2)$ are the "actions".

In this normal form the T^{n+1} action on $\mathcal{U}$, respecting α, and having the same orbits as $\mathcal{X}$ appears naturally as:

$$(\alpha_k, \beta_j) \cdot (\theta_k, q_k, z_j) = (\theta_k + \alpha_k, q_k, e^{i2\pi\beta_j} z_j), \quad k = 0, \cdots, r \; ; \; j = 1, \cdots, n-r.$$

Proof

By Theorem B.2, there is a neighborhood $\mathcal{U}$ of F_{x_0}, saturated with orbits carrying an effective action of T^{n+1} which preserve α and $\mathcal{X}$, and has the same orbits.

The proof will be carried out in several steps.

Step 1: The isotropy of x_0 is connected.

Let $p = n-r$ denote the dimension of the isotropy $\mathcal{I}_{x_0}$ of the point x_0. The identity component of $\mathcal{I}_{x_0}$ is a p-dimensioanal subtorus of T^{n+1}. After reparametrisation, we may write $T^{n+1} = T^{r+1} \times T^p$, where the second factor corresponds to the connected component of $\mathcal{I}_{x_0}$. In fact the isotropy is a product $\Gamma \times T^p$, where Γ is a finite subgroup of T^{r+1}. We are going to show that Γ is the trivial group, hence that the isotropy is connected, meaning also that the factor T^{r+1} acts freely on a neighborhood of F_{x_0}.

Any $\gamma \in \Gamma$ acts by the "slice representation" on the $2n-r$ dimensional tangent space to the slice at x_0 and this action respects the 2-form induced by $d\alpha$ and the 1-forms induced by $i_{Y_i}d\alpha$, where $Y_0, \cdots, Y_r$ form a basis of fundamental vector fields of the T^{r+1} action. Among these 1-forms, there are r linearly independent and it is easy to see that T^p can be identified with a maximal torus preserving them. Since γ commutes with T^p, there exists $\gamma_1 \in T^p$ having the same action as γ on the slice. This implies that $\gamma - \gamma_1$ acts trivially. Since the action is effective, we conclude that $\gamma = e$.

Step 2: After an eventual shrinking of $\mathcal{U}$, we may introduce coordinates
$$(\theta_0, \cdots, \theta_r, u_1, \cdots, u_{2n-r})$$
which identify $\mathcal{U}$ with $T^{r+1} \times \Omega$, where Ω is a neighborhood of 0 in $\mathbb{R}^{2n-r}$, in such a way F_{x_0} is defined by $u = 0$ and that the action of T^{r+1} on $\mathcal{U}$ is the translation on the first factor.

In this situation, the fundamental vector fields of action $Y_0, \cdots, Y_r$ are just $\frac{\partial}{\partial \theta_0}, \cdots, \frac{\partial}{\partial \theta_r}$.

Consider the functions:
$$q_i = i_{\frac{\partial}{\partial \theta_i}} \alpha, \qquad i = 0, \cdots, r.$$

The 1-forms

$$dq_i = -i_{\frac{\partial}{\partial \theta_i}} d\alpha$$

are linearly independent at x_0. Hence, we may assume that the q_i is a subcollection of coordinates taken from $(u_1, \cdots, u_{2n-r})$, after an eventual shrinking of $\mathcal{U}$ and a permutation of the coordinates θ_i. Now our system of coordinates can be written:

$$(\theta_0, \cdots, \theta_r, q_1, \cdots, q_r, v_1, \cdots, v_{2(n-r)})$$

and the contact form α can be written in these coordinates:

$$\alpha = \sum_{i=0}^{r} q_i d\theta_i + \alpha_1$$

where depends only on the variables (v, q). Consequently:
$i_{\frac{\partial}{\partial \theta_i}} \alpha = i_{\frac{\partial}{\partial \theta_i}} d\alpha = 0$.

Step 3: The isotropy T^p acts in the space of variables (q, v), preserving the $q_i's$ and the 2-form $d\alpha_1$: indeed

$$Z \cdot dq_i = L_Z(i_{Y_i} \alpha) = i_{Y_i} L_Z \alpha = 0,$$

and $L_Z \alpha = L_Z \alpha_1 - 0$, for any fundamental vector field of the action of T^{n+1}.

We can linearize the action of T^p so to replace the coordinates $v_1, .., v_{2p}$ by complex coordinates $(z_1, .., z_p)$ in such a way the action is given by:

$$(\beta_1, \cdots, \beta_p) \cdot (q_1, \cdots, q_r, z_1, \cdots, z_p) = (q_1, \cdots, q_r, e^{2i\pi\beta_1} z_1, \cdots, e^{2i\pi\beta_p} z_p).$$

The 2-form $d\alpha_1$ induces an invariant symplectic form on the subspace $\{q_i = 0\}$, and by continuity, a symplectic form with parameters q_i on the affine subspace where q_i are constant.

Using an equivariant method of paths of Moser with parameters [30], we can make the 2-form $d\alpha_1$ assume in polar coordinates the following expression:

$$d\alpha_1 = \sum_{j=1}^{p} \rho_j d\rho_j \wedge d\phi_j$$

modulo $dq_i, \cdots, dq_r$. We have:

$$i_{\frac{\partial}{\partial \phi_j}} d\alpha = i_{\frac{\partial}{\partial \phi_j}} d\alpha_1 = -\rho_j d\rho_j + \sum_{i=1}^{r} f_{ij}(q, z) dq_i.$$

Since $\left(\frac{\partial}{\partial \phi_j}\right)$ are the fundamental vector fields of the action, which preserves α, the 1-form above is closed. As a result, the f_{ij} do not depend on z, and indeed $\sum_{i=1}^{r} f_{ij} dq_i$ is a closed 1-form in variables q_i. Shrinking $\mathcal{U}$, we may assume that:

$$i_{\frac{\partial}{\partial \phi_j}} d\alpha = -\rho_j d\rho_j - d\big(g_j(q)\big)$$

and hence:

$$i_{\frac{\partial}{\partial \phi_j}} \alpha = \frac{1}{2}\rho_j^2 + g_j(q) + C$$

where C is some constant.

On the other hands, when $\rho_i \longrightarrow 0$, the fundamental vector field $\frac{\partial}{\partial \phi_j}$ vanishes. Hence $g_j(q) + C = 0$. Therefore the contact form α assumes the following expression:

$$\alpha = \sum_{i=0}^{r} q_i d\theta_i + \sum_{j=1}^{p} \frac{1}{2}\rho_j^2 d\varphi_j + \beta$$

where the 1-form β contains only the variables q_i and ρ_j, which are first integrals of the action of T^{n+1}. The 1-form β can always be written under this form:

$$\beta = \sum_{i=1}^{r} \beta_i\big(q_1, \cdots, q_r, \rho_1^2, \cdots, \rho_p^2\big) dq_i + \sum_{j=1}^{p} \beta_j'\big(q_i, \cdots, q_r, \rho_1^2, \cdots, \rho_p^2\big) \rho_j d\rho_j.$$

Let us now consider the following change of coordinates:

$$\tilde{\theta}_i = \theta_i - \beta_i, \qquad \tilde{\phi}_i = \phi_i - \beta_j'.$$

We get:

$$\alpha = \sum_{i=0}^{r} q_i \tilde{\theta}_i + \sum_{j=1}^{p} \frac{1}{2}\rho_j^2 d\tilde{\phi}_j + dh$$

where

$$h = \sum_{i=0}^{r} \beta_i q_i + \sum_{j=1}^{p} \frac{1}{2} \beta_j' \rho_j^2$$

is a first integral of the T^{n+1} action, in particular invariant by the Reeb field Z of α. Setting

$$\alpha_0 = \sum_{i=0}^{r} q_i \tilde{\theta}_i + \sum_{j=1}^{p} \frac{1}{2} \rho_j^2 d\tilde{\varphi}_j$$

we have: $\alpha = \alpha_0 + dh$, with $Z \cdot h = 0$. Let $\alpha_t = \alpha_0 + tdh$. The flow ψ_t of $X = -hZ$ commutes with the T^{n+1} action, we have, as in the proof of Theorem B.1:

$$\frac{d}{dt}(\psi_t^* \alpha_t) = \psi_t^* \left[L_X \alpha_t + \frac{d}{dt} \alpha_t \right] = 0.$$

Hence $\psi_1^* \alpha = \alpha_0$, which allows us to find coordinates in which the form α has the normal form we wanted. The fact that q_0 depends only on $q_i's$ and ρ_j^2 is a consequence of the fact it is invariant by the action. $\qquad\square$

Let us now give an explicit expression of q_0:

Let $(P, H, \mathcal{X}, \mathcal{A})$ be the CIC-structure on the compact $2n+1$ dimensional manifold P with an oriented contact structure H and let α be the contact form obtained in Theorem 2, which is invariant by $\mathcal{X}$ and by all the T^{n+1}-local actions (see Theorem 2). Denote by W the orbits space of the action of $\mathcal{X}$ and by $\pi : P \longrightarrow W$ the canonical projection. We endow W with the quotient topology. By Theorem 2, W is Hausdorff, and it is compact since P is.

Let F_{x_0} be an $r + 1$ dimensional orbit of $\mathcal{X}$ through the point $x_0 \in P$. Then $F_{x_0} = \pi^{-1}(y_0)$ where $y_0 = \pi(x_0)$.

Let $\mathcal{U}$ be the neighborhood of F_{x_0} which is the domain of the contact singular angle-action coordinates $(\theta, q, \phi, \frac{1}{2}\rho^2)$ constructed above, and let $\widetilde{\mathcal{U}} = \pi(\mathcal{U})$. The functions

$$q_1, \cdots, q_r, \frac{1}{2}\rho_1^2, \cdots, \frac{1}{2}\rho_p^2$$

are π-projectable, and hence may be viewed as defined on $\widetilde{\mathcal{U}}$. They define a homeomorphism

$$\phi_{\widetilde{\mathcal{U}}} : \widetilde{\mathcal{U}} \longrightarrow \widetilde{\Omega}$$

where $\tilde{\Omega}$ is an open subset of $\mathbb{R}^r \times \mathbb{R}_+^p$.

In the open set $\mathcal{U}$, the Reeb field Z of the contact form α can be written:

$$Z = \sum_{i=0}^{r} Z_i(q, \rho^2) \frac{\partial}{\partial \theta_i} + \sum_{j=0}^{p} Z_i'(q, \rho^2) \frac{\partial}{\partial \phi_j}.$$

Note that the coefficients Z_i, Z_j' of Z are first integrals of the action and hence must be functions of q_i's and ρ^2's alone.

Since

$$\alpha = \sum_{i=0}^{r} q_i d\theta_i + \sum_{j=1}^{p} \frac{1}{2} \rho_j^2 d\phi_j$$

the equation $\alpha(Z) = 1$ gives:

$$\sum_{i=0}^{r} Z_i(q, \rho^2) q_i + \frac{1}{2} \sum_{j=0}^{p} Z_i'(q, \rho^2) \rho_j^2 = 1.$$

Writing that $i(Z)d\alpha = 0$, we get:

$$0 = \sum_{i=0}^{r} Z_i(q, \rho^2) dq_i + \sum_{j=0}^{p} Z_i'(q, \rho^2) \, \rho_j \, d\rho_j.$$

Hence at x_0,

$$(i_Z d\alpha)(x_0) = \sum_{i=0}^{r} Z_i(q, 0) dq_i$$

which gives:

$$Z_0(q, 0) dq_0 = -\sum_{i=1}^{r} Z_i(q, 0) dq_i.$$

If $Z_0(q, 0) = 0$, then $\displaystyle\sum_{i=0}^{r} Z_i(q, 0) dq_i = 0$.

Since $\{dq_i\}_{1, \cdots, r}$ are linearly independent, we conclude that $Z_i(q, 0) = 0$ for $i = 1, \cdots, r$. This contradict the fact that $\displaystyle\sum_{i=0}^{r} Z_i(q, 0) q_i = 1$.

Therefore $Z_0(q, 0) \neq 0$, and hence Z_0 will be non-zero on a neighborhood of x_0, we may assume to be our neigbhorhood (after shrinking it eventually). Hence we can solve for q_0 and write:

$$q_0 = \frac{1}{Z_0(q, \rho^2)} - \sum_{i=1}^{r} \frac{Z_i(q, \rho^2)}{Z_0(q, \rho^2)} q_i - \frac{1}{2} \sum_{j=1}^{p} \frac{Z_j'(q, \rho^2)}{Z_0(q, \rho^2)} \rho_j^2.$$

Thus we see that q_0 is a smooth function of q_i's, $i = 1, \cdots, r$, and of ρ's.

B.4 The manifold of invariant Tori

B.4.1 Admissible change of action coordinates

Since the local coordinates are defined by:

$$q_i = \alpha\left(\frac{\partial}{\partial\theta_i}\right)$$

$$\frac{1}{2}\rho_j^2 = \alpha\left(\frac{\partial}{\partial\phi_j}\right) \tag{B.4.1}$$

the allowed new coordinates will be

$$q_i' = \alpha\left(\frac{\partial}{\partial\theta_i'}\right)$$

$$\frac{1}{2}(\rho_j')^2 = \alpha\left(\frac{\partial}{\partial\phi_j'}\right) \tag{B.4.2}$$

where $\alpha(\frac{\partial}{\partial\theta_i'})$, $\alpha(\frac{\partial}{\partial\phi_j'})$ are a change in the basis of the Lie algebra of T^{n+1}, preserving the isotropy subgroup (the strata of the action defined by the dimension of the orbits). Hence these changes of the basis must necessarily be of these forms:

$$\frac{\partial}{\partial\phi_j'} = \frac{\partial}{\partial\phi_{\sigma(j)}}$$

$$\frac{\partial}{\partial\theta_i'} = \sum_{k=0}^{r} A_k^i \frac{\partial}{\partial\theta_k} + \sum_{l=1}^{p} B_l^i \frac{\partial}{\partial\phi_l} \tag{B.4.3}$$

where σ is a permutation of $\{1, \cdots, p\}$, and A_k^i, B_l^i are integers. Evaluating these vector fields on α, and using the above expression of q_0 we get:

$$(\rho_j')^2 = (\rho_j)^2$$

$$q_i' = \frac{A_0^i}{Z_0} + \sum_{k=1}^{r}\left[A_k^i - A_0^i\frac{Z_k}{Z_0}\right]q_k + \sum_{j=1}^{p}\frac{1}{2}\left[B_j^i - A_0^i\frac{Z_j'}{Z_0}\right]\rho^2.$$

These change of coordinates are obviously smooth. Hence the action coordinates define on W a structure of a smooth n dimensional manifold with boundary and corners modelled on $\mathbb{R}^r \times \mathbb{R}_+^p$.

B.4.2   Legendre lattices

Following Duistermaat [16], we are going to introduce the notion of "Legendre Lattice" in the study of the manifold W. We first analyze the regular case $n = r$ or $p = 0$.

The regular case

Consider a smooth n-dimensional manifold without boundary X and let $J^1 X$ denote the manifold of 1-jets of smooth functions on X, equipped with its canonical contact form α_1. Denote by $\pi_1 : J^1 X \longrightarrow X$ the natural fibration.

Definition B.5

A legendre lattice on X is a Legendre submanifold $\mathcal{R}_1$ of $(J^1 X, \alpha_1)$ such that:

1. the restriction of π_1 to $\mathcal{R}_1$ is a covering map $\pi_1 : \mathcal{R}_1 \longrightarrow X$

2. for each fiber $J^1_x X$ over $x \in X$, $\mathcal{R}_1 \cap J^1_x X$ is a lattice in the $(n+1)$-dimensional vector space $J^1_x X$.

Let P^1 be the total space of the bundle over X whose fiber over $x \in X$ is $J^1_x X / (\mathcal{R}_1)_x$. Then the fibers of the fibration $\hat{\pi} : P^1 \longrightarrow X$ are (n+1) tori.

Over a trivializing open set $U \subset X$ for the bundle $\pi_1 : J^1 X \longrightarrow X$, the lattice $\mathcal{R}_1$ will be an (integral) combination of 1-jets $\{j^1 q_i\}$ of (n+1) functions $q_0, q_1, \cdots, q_n : U \longrightarrow \mathbb{R}$, we call **a local basis** of $\mathcal{R}_1$.

Definition B.6

The coefficients of the lattice $\mathcal{R}_1$ with respect to the local basis $(q_0, \cdots, q_n : U \longrightarrow \mathbb{R})$ are the functions $C_0, \cdots, C_n : U \longrightarrow \mathbb{R}$ defined by

$$j^1 1 = \sum_{i=0}^{n} C_i j^1 q_i.$$

By definition of the 1-jets:

$$\sum_{i=0}^{n} C_i q_i = 1 \qquad\qquad (B.4.4)$$

$$\sum_{i=0}^{n} C_i dq_i = 0. \qquad\qquad (B.4.5)$$

By shrinking U, we may assume that $q_1, \cdots, q_n$ form a local coordinate system on U. We say that the corresponding chart is **adapted** to the lattice. Since $\sum_{i=1}^{n} C_i dp_i = -C_0 dq_0$, and since $dq_1, \cdots, dq_n$ are linearly indepen- dent, $C_0 = 0$ would imply that all the other C_i are zero, contradicting (8). Therefore on U, $C_0 \neq 0$ and then we can solve for q_0 and write:

$$q_0 = \frac{1}{C_0} - \sum_{i=1}^{n} \left(\frac{C_i}{C_0} \right) q_i.$$

Suppose now X is the manifold of invariant tori of a completely inte- grable contact structure without singularities and let $q_i = \alpha(\frac{\partial}{\partial \theta_i})$, then the linear combinations over the integers of dq_i's is a Lagrangian lattice over U. The change of coordinates over $U \cap U'$, shows that dq_i''s are linear combi- nations of dq_i's with integer coefficients. This then implies that the lattices over U and over U' with bases $q_0, \cdots, q_n$ and $q_0', \cdots, q_n'$ agree on $U \cap U'$. Thus we get the following.

Proposition B.7

If X is an integral affine manifold, $J^1 X$ carries a natural Legendre lattice.

The General Case

We consider now a smooth n-dimensional manifold X modeled on $\mathbb{R}^r \times \mathbb{R}^p_+$, where $p = n - r$ takes values from 0 to n. This is a so-called "manifold with boundary and corners". The smooth functions on a chart of X are restrictions of smooth functions on $\mathbb{R}^r \times \mathbb{R}^p$. The manifold $J^1 X$ of 1-jets is $(2n + 1)$ dimensional manifold with boundary and corners, with a natural contact form α_1. We still denote by $\pi_1 : J^1 X \longrightarrow X$ the natural vector bundle over X.

Definition B.7

A Legendre lattice on X is a Legendre submanifold with boundary and corners $\mathcal{R}_1$ of $(J^1 X, \alpha_1)$ such that:

1. $\pi_1 : \mathcal{R}_1 \longrightarrow X$ is a covering map

2. the trace of $\mathcal{R}_1$ on each fiber of $J^1 X$ is a lattice in that fiber

3. each point of X has an open neighborhood U with local coordinates
$(q_1, \cdots, q_n)$ *so that*
$j^1 q_0, j^1 q_1, \cdots, j^1 q_n$ *is a local basis of $\mathcal{R}_1$, where q_0 is some differentiable function of $(q_1, \cdots, q_n)$.*

(We say that this is an adapted local basis for the lattice.)

We saw that in the regular case, condition (3) was automatically satisfied. Let us observe also that in a neighborhood of a point belonging to an r-dimensional face of X, the change of charts which are adapted to the lattice necessary have the form: $q'_j = q_{\tau(j)}$ for $j \geqslant r$ where τ is a permutation of $(r+1, \cdots, n)$, since the only elements of $\mathcal{R}_1$ consisting into jets of functions vanishing on the hyperplane $q_j = 0$ are integer multiples of $j^1 q_j$.

The coefficients of the lattice in a local basis are defined by the same formulas.

For each $y \in X$, denote by $\widehat{\mathcal{R}}_{1y}$ the abelian subgroup of $J_y^1 X$ generated by the fiber $\mathcal{R}_{1y}$ of the lattice and the 1-jets of all functions which vanish identically on the face of X to which belongs y. In particular if y is not a boundary point, then $\widehat{\mathcal{R}}_{1y} = \mathcal{R}_{1y}$. Define now:

$$P_y^1 = J_y^1 X / \widehat{\mathcal{R}}_{1y}, \qquad P^1 = \bigcup_{y \in X} P_y^1.$$

Let $\{j^1 q_0, \cdots, j^1 q_n\}$ be an adapted local basis of $\mathcal{R}_1$ in the open set U. Suppose that the most degenerate stratum in the boundary of X which meets U has dimension r, i.e. the chart $(q_1, \cdots, q_n)$ has its values in $\mathbb{R}^r \times \mathbb{R}_+^p$. Since any element of $J^1 U$ over $y \in U$, with coordinates $(q_i, \cdots, q_n)$ can be written as: $\sum_{i=0}^{n} p_i \left(j_y^1(q_i) \right)$, it follows that we get $(q_1, \cdots, q_n, p_0, \cdots, p_n)$ as coordinates on $J^1 U$. The correspondence:

$$(q_1, \cdots, q_n, p_0, \cdots, p_n)$$

$$\downdownarrows$$

$$\left(e^{(2i\pi)p_0}, \cdots, e^{(2i\pi)p_r}, q_1, \cdots, q_r, (2q_{r+1})^{1/2} e^{(2i\pi)p_{r+1}}, \cdots, (2q_n)^{1/2} e^{(2i\pi)p_n} \right)$$

factors through a local chart for P^1 with values in an open set of $T^{r+1} \times \mathbb{R}^r \times \mathbb{C}^p$. These charts determine on P^1 a structure of smooth manifold with boundary and corners. The natural projection $\hat{\pi}_1 : P^1 \longrightarrow X$ appears as a singular fibration with fibers tori of different dimensions ranging from zero to $n + 1$; the neutral elements in each fiber come from the projection of 1-jets of the zero function on X from J^X to P^1. Moreover the canonical contact form on $J^1 X$ descends to a contact form α_1 on P^1.

The Legendre Lattice of W

We have shown that the manifold of invariant tori W is an n-dimensional manifold with boundary and corners. Let U be an open neighborhood of an orbit with singular contact angle action coordinates:

$$\left(\theta_0, \cdots, \theta_r, q_1, \cdots, q_r, \phi_i, \cdots, \phi_p, \frac{1}{2}\rho_1^2, \cdots, \frac{1}{2}\rho_p^2\right).$$

The functions

$$q_1, \cdots, q_r, q_{r+1} = \frac{1}{2}\rho_1^2, \cdots, q_n = \frac{1}{2}\rho_p^2$$

can be considered as coordinates on $\tilde{U} = \pi(U)$. Recall that $q_0 = \alpha\left(\frac{\partial}{\partial\theta_0}\right)$. The linear combinations of $\{j^1 q_0, \cdots, j^1 q_n\}$ with integer coefficients is a Legendre lattice over $\tilde{U}$. This lattice is invariant under change of singular contact angle action coordinates. We saw that the change in $q's$ is affine, the linear part of which being represented by a matrix with integer entries. The lattice $\mathcal{R}_1$ is therefore invariantly well defined. Looking back at equations (3) and (10), we see that the local coefficients for this lattice are the components of the Reeb field, considered as functions on $\tilde{U}$.

We summarize the results of this section in the following.

Theorem B.4

Let $(P, H, \mathcal{X}, \mathcal{A})$ be a CIC-structure on the compact $2n + 1$ dimensional manifold P with an oriented contact structure H. Let $\pi : P \longrightarrow W$ denote the canonical projection onto the orbit space of $\mathcal{X}$. Then

1. *W has a natural structure of manifold with boundary and corners.*

2. *W carries a natural Legendre lattice $\mathcal{R}_1$.*

3. *For each open set $U = (\pi)^{-1}(\tilde{U})$ with SCAAC*

$$\left(\theta_0, \cdots, \theta_r, q_1, \cdots, q_r, \phi_1, \cdots, \phi_p, \frac{1}{2}\rho_1^2, \cdots, \frac{1}{2}\rho_p^2\right)$$

the 1-jets of the functions defined in $\tilde{U}$ by:

$$q_0 = \alpha \left(\frac{\partial \theta_0}{\partial} \right), q_1, \cdots, q_r, q_{r+1} = \frac{1}{2}\rho_1^2, \cdots, q_n = \frac{1}{2}\rho_p^2$$

form a local basis for the lattice $\mathcal{R}_1$ in which the local coefficients of the lattice are the components of the Reeb field of the contact form α (obtained in Theorem B.2, which is invariant by $\mathcal{X}$ and by all the T^{n+1}-local actions (see Theorem B.2)).

In the regular case, π is a local trivial fibration with fibers tori T^{n+1}: the situation is then the exact equivalent in contact geometry of Duistermaat lagrangian fibrations [16].

The singular case is the contact analog of the "singular lagrangian fibrations" of Boucetta-Molino [10].

B.4.3   The Chern class of the singular fibration π : $P \longrightarrow W$

Consider an open cover $(\tilde{U}_a)_{a \in A}$ of W, such that for each $a \in A$, the open set $U_a = \pi^{-1}(\tilde{U}_a)$ is the domain of singular contact angle-action coordinates (SCAAC):

$$\left(\theta_0^a, \cdots, \theta_r^a, q_1^a, \cdots, q_r^a, \phi_1^a, \cdots, \phi_p^a, \frac{1}{2}(\rho_1^a)^2, \cdots, \frac{1}{2}(\rho_p^a)^2 \right).$$

For simplicity, we will write:

$$\begin{cases} q_{r+j}^a &= \frac{1}{2}(\rho_j^a)^2, \\ \theta_{r+j}^a &= \phi_j^a, \qquad j = 1, \cdots, p \end{cases} \tag{B.4.6}$$

and recall that $q_0 = \alpha \left(\frac{\partial}{\partial \theta_0} \right)$ and that finally in U_a,

$$\alpha = \sum_{i=0}^{n} q_i^a d\theta_i^a.$$

We have:

On the intersection $U_a \cap U_b$, we gave the following transition formulas:

$$q_i^b = \sum_{j=0}^{n} k_{ij}^{ab} q_j^a \tag{B.4.7}$$

$$\theta_i^b = \sum_{j=0}^{n+1} m_{ij}^{ab}(\theta_j^a + f_j^{ab}) \tag{B.4.8}$$

where f_j^{ab} are functions of $q's$ satisfying: $\sum_{i=0}^{n} q_i^a df_i^{ab} = 0$ where

$$(k_{ij}^{ab}), (m_{ij}^{ab}) \in SL(n+1, \mathbb{R})$$

are inverse of each another.

Proof

The proof is similar to the proof of the analog fact in symplectic geometry. For completeness, we give it here. Since on $U_a \cap U_b$,

$$\Omega = \sum_{j=0}^{n} dq_j^a \wedge d\theta_j^a = \sum_{j=0}^{n} dq_j^b \wedge d\theta_j^b.$$

For each $j = 0, 1, \cdots, n$, we have:

$$i_{\frac{\partial}{\partial \theta_j^b}} \Omega = dq_j^b = \sum_{i=1}^{n} \left(\frac{\partial}{\partial q_i^a} q_j^b \right) dq_i^a.$$

We also have:

$$i_{\left(\sum_{l=1}^{n} \frac{\partial q^{b_j}}{\partial q^{a_l}} \frac{\partial}{\partial \theta^{a_l}} \right)} \Omega = \left(\sum_{l=1}^{n} \frac{\partial q^{b_j}}{\partial q^{a_l}} i_{\frac{\partial}{\partial \theta^{a_l}}} \right) \left(\sum_{j=0}^{n} dq_j^a \wedge d\theta_j^a \right)$$

$$= \sum_{l=1}^{n} \left(\frac{\partial q^{b_l}}{\partial q^{a_l}} \right) dq^{a_l} = dq^{b_j}$$

$$= i_{\frac{\partial}{\partial \theta^{b_j}}} \Omega. \tag{B.4.9}$$

Hence

$$\frac{\partial}{\partial \theta_j^b} = \sum_{l=0}^{n} \left(\frac{\partial q_j^b}{\partial q_l^a} \right) \frac{\partial}{\partial \theta_l^a}.$$

Therefore for fixed $q \in W$, the vector fields $\left\{ \frac{\partial}{\partial \theta_j^b} \right\}$ are linear combinations

of $\left\{ \frac{\partial}{\partial \theta_i^a} \right\}$. This means that for fixed q,

$$\theta_j^b = \sum A_{ij} \theta_i^a + t_j$$

where A_{ij} and t_j are real numbers depending on q. Since θ_j^a, θ_l^b are 2π-periodic, these numbers must be integers. But on the other hands $A_{ij} = A_{ij}(q) = \left(\frac{\partial q_j^b}{\partial q_i^a} \right)(q^a)$ is smooth in q; therefore it must be a constant, we denote k_{ij}^{ab}. This implies that:

$$q_i^b = \sum_j k_{ij}^{ab} q_j^a.$$

Now:

$$\sum_{j=0}^{n} dq_j^b \wedge d\theta_j^b = \sum_{ij} k_{ij}^{ab} dq_j^a \wedge d\theta_j^b = \sum_{j=0}^{n} dq_j^a \wedge d\theta_j^a$$

i.e.

$$d\theta_j^a = \sum_i k_{ij} d\theta_i^b + w_j dq_j^a$$

where w_j are functions of $q's$. Since

$$d(w_j dq_j^a) = d(d\theta_j^a) - d\left(\sum_i k_{ij} \theta_i^b \right) = 0,$$

then $w_j dq_j^a = d(-f_j^{ab})$ by Poincare lemma. For we can assume that $U_a \cap U_b$ is contractible, provide that the cover is fine enough to be geodesically convex for some riemannian metric. Hence if (m_{ij}) is the inverse of the matrix (k_{ij}), we see that

$$\theta_j^b = \sum m_{ij} (\theta_i^a + f_i^{ab}).$$

The fact that

$$\sum_{j=0}^{n} q_j^a \wedge d\theta_j^a = \sum_{j=0}^{n} q_j^b \wedge d\theta_j^b$$

and the formula above, implies that

$$\sum_{l=0}^{n} q_l^a df_l^{ab} = 0.$$

The proof of the proposition is complete. □

Consider now the following section of the 1-jet bundle J^1W:

$$\phi_{ab} = \sum_{i=0}^{n} f_i^{ab} j^1 q_i^a.$$

This section is defined modulo the lattice $\mathcal{R}_1$ since the functions f_i^{ab} are defined modulo integer constant. Hence, its projection $\overline{\phi}_{ab}$ is a well defined section of $J^1W/\mathcal{R}_1$.

Furthermore let $F^{ab} = (f_0^{ab}, \cdots, f_n^{ab})$ and $\theta^u = (\theta_0^u, \cdots, \theta_n^u)$, then $F^{ab} = K\theta^b - \theta^a$, where K is the matrix (k_{ij}) above. Hence:

$$
\begin{aligned}
\phi_{ab} &= \sum_i \left(\sum_j k_{ij}^{ab} \theta_j^b - \theta_i^a \right) j^1 q_i^a = -\sum_i \theta_i^a j^1 q_i^a + \sum_{ij} k_{ij} j^1 q_i \theta_j^b \\
&= -\sum_i \theta_i^a j^1 q_i^a + \sum_k \theta_k^b j^1 q_k^b.
\end{aligned}
\tag{B.4.10}
$$

This formula proves that

$$\overline{\phi}_{ab} + \overline{\phi}_{bc} + \overline{\phi}_{ca} = 0.$$

On $U_a \cap U_b \cap U_c$. Therefore $\underline{\phi}_{ab}$ is a well defined cocycle on W with values in the sheaf $\underline{J^1W/\mathcal{R}_1}$ of sections of the bundle $J^1W/\mathcal{R}_1$. Denote by:

$$[\phi] \in H^1\left(W, \underline{J^1W/\mathcal{R}_1}\right)$$

its cohomology class.

Consider the following exact sequence of sheaves:

$$0 \longrightarrow \underline{\mathcal{R}_1} \longrightarrow \underline{J^1W} \longrightarrow \underline{J^1W/\mathcal{R}_1} \longrightarrow 0.$$

Since $\underline{J^1W}$ is a fine sheaf, the coboundary operator is an isomorphism

$$H^1(W, \underline{J^1W/\mathcal{R}_1}) \approx H^2(W, \mathcal{R}_1).$$

The image $\gamma \in H^2(W, \mathcal{R}_1)$ of $[\phi]$ is called the Chern class of the singular fibration $\pi : P \longrightarrow W$. It can be represented by the cocycle:

$$\gamma_{abc} = \phi_{ab} + \phi_{bc} + \phi_{ca}.$$

Remark B.2

The section ϕ_{ab} is a holonomic section, i.e.

$$\phi_{ab} = j^1 \left(\sum_{i=0}^{n} f_i^{ab} q_i^a \right)$$

since $\sum q_i^a df_i^{ab} = 0$.

Observe that to the contrary of what happens in Duistermaat lagrangian fibrations, this fact does not carry a constraint on the Chern class since we have the following:

Proposition B.8

Any 1-cocycle with values in $\underline{J^1 W}/\mathcal{R}_1$ is cohomologous to a cocycle defined by local holonomic sections.

Proof

Let $J^0 W = W \times \mathbb{R}$ and $p_0 : J^1 W \longrightarrow J^0 W$ be the natural projection. As above, we denote by $\underline{J^k W}$ the sheaf of germs of local sections of $J^k W$ and let $\mathcal{R} \subset \underline{J^0 W}$ be the subsheaf of germs of functions whose 1-jets belong to $\mathcal{R}_1$. We have the following exact sequence of sheaves:

$$0 \longrightarrow \underline{J^0 W}/\mathcal{R} \overset{j^1}{\longrightarrow} \underline{J^1 W}/\mathcal{R}_1 \overset{D}{\longrightarrow} \underline{T^* W} \longrightarrow 0.$$

Here $\underline{T^* W}$ is the sheaf of germs of 1-forms on W and D is the Spencer operator $D(\sigma) = \sigma - j^1(p_0 \circ \sigma)$ [38], [39]. Since $\underline{T^* W}$ is a fine sheaf, the cohomology exact sequence of the exact sequence above yields the following isomorphism:

$$H^1(W, \underline{J^0 W}/\mathcal{R}_0) \approx H^1(W, \underline{J^1 W}/\mathcal{R}_1).$$

□

Remark B.3

The Chern class $[\gamma]$ is independent of the coice of the particular singular action-angles coordinates on the open cover $\{U_a\}$, and of this open cover. A change of these objects gives rise to a cocycle which is cohomologous to the old one.

B.4.4 The classification theorem

We summarize what we did so far: with each completely integrable contact structure $(P, H, \mathcal{X}, \mathcal{A})$, we associated natural objects:

1. The canoincal projection $\pi : P \longrightarrow W$ on the orbits space W of $\mathcal{X}$.

2. A Legendre Lattice $\mathcal{R}_1$ on W.

3. A Chern class $[\gamma] \in H^2(W, \mathcal{R}_1)$.

The triple $(W, \mathcal{R}_1, [\gamma])$ is called the "**characteristic invariants**" of the singular fibration $\pi : P \longrightarrow W$. The goal of the next section is to study how to extend the characteristc invariants determine the singular foliation.

Proposition B.9
   If Z and Z' are the Reeb fields of α and of α', then $\Phi_* Z = Z'$.

   Proof
   It is clear that $(\Phi^{-1})_* Z'$ is the Reeb field of $\Phi^* \alpha'$, since

$$\begin{cases} i_{(\Phi^{-1})_* Z'}(\Phi^* \alpha') & = & 1 \\ i_{(\Phi^{-1})_* Z'}(d\Phi^* \alpha') & = & 0. \end{cases} \tag{B.4.11}$$

Since

$$Z = \sum_{i=0}^{r} Z_i \frac{\partial}{\partial \theta_i} + \sum_{j=1}^{n-r} Z'_j \frac{\partial}{\partial \phi_j}$$

where θ and ϕ are the angle coordinates and

$$\beta = \sum_{i=1}^{n} \beta_i(q) dq_i$$

where $q = (q_1, \cdots, q_n)$ are the action coordinates, we have:

$$i_Z \beta = i_Z d\beta = 0.$$

Hence: $i_Z(\alpha - \pi^* \beta) = 1$ and $i_Z(d\alpha - \pi^* d\beta) = 0$.
   This means that Z is the Reeb field of $\Phi^* \alpha' = \alpha - \pi^* \beta$. By uniqueness of the Reeb field, $Z = (\Phi^{-1})_* Z'$. □

Theorem B

1. *Let W be an n-dimensional compact manifold with boundary and corners, let $\mathcal{R}_1$ be a Legendre lattice on W and let $[\gamma] \in H^2(W, \mathcal{R}_1)$. Then there exists a completely integrable contact structure $(P, H, \mathcal{X}, \mathcal{A})$ with characteristic invariants $(W, \mathcal{R}_1, [\gamma])$.*

2. *If two completely integrable contact structure $(P, H, \mathcal{X}, \mathcal{A})$ and $(P', H', \mathcal{X}', \mathcal{A}')$ have the same characteristic invariants, then there exists a contact diffeomorphism $\Phi : P \longrightarrow P'$ such that $\pi = \pi' \circ \Phi$ where π, π' are the canonical projections $P \longrightarrow W$ and $P' \longrightarrow W$.*

Proof

(i) Existence:

Let $(\tilde{U}_a)_{a \in A}$ be a Leray open cover of W and for each $a \in A$, a calibrated basis $\{j^1 q_0^a, \cdots, j^1 q_n^a\}$ for the lattice $\mathcal{R}_1$ over $\tilde{U}_a$.

On $\tilde{U}_{ab} = \tilde{U}_a \cap \tilde{U}_b$, we have the change of coordinates:

$$q_i^b = \sum_{j=0}^{n} K_{ij}^{ab} q_j^a$$

where the matrix (K_{ij}^{ab}) has the form imposed by the condition for the basis of the lattice to be adapted.

By Proposition 7, the class $[\gamma]$ can be defined through holonomic sections $\psi_{ab} : \tilde{U}_{ab} \longrightarrow J^1 W$. We thus have:

$$\psi_{ab} = j^1 f^{ab} = \sum_{i=0}^{n} f_i^{ab} j^1 q_i^a.$$

Hence $f^{ab} = \sum_{i=0}^{n} f_i^{ab} q_i^a$ and $\sum_{i=0}^{n} q_i^a df_i^{ab} = 0$.

On $\overline{U}_a = \pi^{-1}(\tilde{U}_a) \subset J^1 W$, we have natural Darboux coordinates $(q^a, p^a) = (q_1^a, \cdots, q_n^a, p_0^a, \cdots, p_n^a)$ in which the canonical contact form α_1 of $J^1 W$ assumes the expression:

$$\sum_{i=0}^{n} q_i^a dp_i^a \tag{B.4.12}$$

where q_0^a is given by the formula in subsection B.4.1. On $\overline{U}_{ab} = \overline{U}_a \cap \overline{U}_b$, we consider the following diffeomorphisms: $\overline{\Psi}_{ab} : \overline{U}_{ab} \longrightarrow \overline{U}_{ab}$ defined by:

$$\overline{\Psi}_{ab}(q^a, p^a) = \left(\sum_{j=0}^{n} K_{ij}^{ab} q_j^a, \sum_{j=0}^{n} M_{ij}^{ab} \left(p_j^a + f_j^{ab} \right) \right). \qquad \text{(B.4.13)}$$

The formula (B.4.13) shows that these diffeomorphisms preserve the contact form α_1.

Recall that in subsection B.4.3 we defined a smooth manifold without boundary P^1 which carries a natural contact form ω which naturally projects onto W so that $\pi_1 : P^1 \longrightarrow W$ is a singular fibration with tori of various dimensions as fibers. We also had a natural projection $\mathcal{P} : J^1 W \longrightarrow P^1$.

If $U_a = \mathcal{P}^{-1}(\tilde{U}_a) = \mathcal{P}(\overline{U}_a)$ and $U_{ab} = U_a \cap U_b$, then $\overline{\Psi}_{ab}$ projects into a contact diffeomorphism:

$$\Psi_{ab} : U_{ab} \longrightarrow U_{ab}.$$

Let (P, α) be the contact compact manifold without boundary obtained by gluing together the disjoint contact manifolds (U_a, ω) via the contact diffeomorphisms Ψ_{ab}. Let $H \subset TP$ be the contact structure defined by α, i.e. $H = \ker \alpha$ and if $\pi : P \longrightarrow W$ is the canonical projection, let $\mathcal{A} = \pi^* C^\infty(W)$ and let $\mathcal{X}$ be the abelian Lie algebra of contact hamiltonian vector fields of elements of $\mathcal{A}$, then the completely integrable contact manifold $(P, H, \mathcal{X}, \mathcal{A})$ has $(W, \mathcal{R}_1, [\gamma])$ as characteristic invariants.

(ii) Classification:

Let $(P, H, \mathcal{X}, \mathcal{A})$ and $(P', H', \mathcal{X}', \mathcal{A}')$ be two completely integrable contact structures on P and P' with the same characteristic invariants. Let π, π' be corresponding fibrations $P \longrightarrow W$ and $P' \longrightarrow W$.

Choose now a Leray cover $\{\tilde{U}_a\}$ of W over which there are singular contact angle action coordinates for $(P, H, \mathcal{X}, \mathcal{A})$ and $(P', H', \mathcal{X}', \mathcal{A}')$. Consider the holonomic sections (see subsection B.4.4)

$$\phi_{ab} = j^1 f_{ab} : \tilde{U}_{ab} = \tilde{U}_a \cap \tilde{U}_b \longrightarrow J^1 W$$

$$\phi'_{ab} = j^1 f'_{ab} : \tilde{U}_{ab} \longrightarrow J^1 W$$

where f_{ab} and f'_{ab} come from the transition maps between local angle-action coordinates.

The 1-cocycles $(\overline{\phi}_{ab})$ and $(\overline{\phi}'_{ab})$ with values in $\underline{J^1W}/\mathcal{R}_1$ are cohomologous by hypothesis: hence there exist sections $\sigma^a : \tilde{U}_a \longrightarrow J^1 W$ such that:

$$j^1 f'_{ab} = j^1 f_{ab} + \sigma^a - \sigma^b \qquad (\text{B.4.14})$$

modulo $\mathcal{R}_1$, in $\tilde{U}_{ab}$.

Let $J^0 W = W \times \mathbb{R}$ and let $\mathcal{R} \subset \underline{J^0W}$ the subsheaf of germs of functions on W whose 1-jets belong to $\mathcal{R}_1$. According to the exact sequence of sheaves:

$$0 \longrightarrow \underline{J^0W}/\mathcal{R} \xrightarrow{\ j^1\ } \underline{J^1W}/\mathcal{R}_1 \xrightarrow{\ D\ } \underline{T^*W} \longrightarrow 0$$

we have that in $\tilde{U}_{ab}$, $D(\sigma^a - \sigma^b) = 0$.

This means that the local 1-forms $D\sigma^a$ fit together into a global 1-form $(-\beta)$ on W. If $\sigma^a = \displaystyle\sum_{i=0}^{n} \sigma_i^a j^1 q_i^a$, then:

$$-\beta = \sum_{i=0}^{n} j^1 q_i^a - j^1 \left(\sum_{i=0}^{n} \sigma_i^a q_i^a \right) = -\sum_{i=0}^{n} q_i^a d\sigma_i^a. \qquad (\text{B.4.15})$$

If $\overline{\pi} : P \longrightarrow W$ and $\overline{\pi}' : P' \longrightarrow W$ are the projections of the contact manifolds P and P' to W, we let $U_a = (\overline{\pi})^{-1}(\tilde{U}_a)$ and $U'_a = (\overline{\pi}')^{-1}(\tilde{U}_a)$. On these open sets we define the following diffeomorphisms:

$$\Phi_a : U_a \longrightarrow U'_a : (q_i^a, \theta_i^a) \longmapsto (q_i^a, \theta_i^a - \sigma_i^a)$$

where (q_i^a, θ_i^a) are singular contact angle actions coordinates in U_a.

According to formulas (B.4.13), these local diffeomorphisms fit together into a global diffeomorphism $\Phi : P \longrightarrow P'$. It is easy to check that $\Phi^* \alpha' = \alpha - \pi^* \beta$ and $\pi' \circ \Phi = \pi$.

To simplify notations, let us write: $\Phi^* \alpha' = \alpha + \omega$ with $\omega = -\pi^* \beta$. We have: $i_Z \omega = i_Z d\omega = 0$.

Consider:

$$\alpha_t + \alpha + t\omega.$$

We have:

$$\alpha_t \wedge (d\alpha_t)^n = \sum_{k=0}^{n} \binom{k}{n} t^k \big(\alpha \wedge (d\omega)^k \wedge (d\alpha)^{n-k}\big)$$

$$+ \sum_{k=0}^{n} \binom{k}{n} t^{k+1} \big(\omega \wedge (d\omega)^k \wedge (d\alpha)^{n-k}\big).$$

For any $x \in P$, consider a basis $\{Z_x, X_1, \cdots, X_n, Y_1, \cdots, Y_n\}$ of $T_x P$, where Z_x is the value at x of the Reeb field Z and $(X_1, \cdots, X_n, Y_1, \cdots, Y_n)$ is a basis of $H_x = \ker \alpha_x$, denoted $(Z, X, Y)_x$ for short, such that

$$\big(\alpha \wedge (d\alpha)^n\big)(Z, X, Y)_x = 1.$$

We have that:

$$\big(\omega \wedge (d\omega)^k \wedge (d\alpha)^{n-k}\big)(Z, X, Y)_x = 0$$

since Z is in the kernel of ω, $d\omega$ and $d\alpha$.

On the other hands, for the same reasons:

$$\big(\alpha \wedge (d\omega)^k \wedge (d\alpha)^{n-k}\big)(Z, X, Y)_x = (d\omega)^k \wedge (d\alpha)^{n-k})(X, Y)_x.$$

Hence we need to evaluate $(d\omega)^k \wedge (d\alpha)^{n-k})$ on the horizontal distribution H.

If any element ξ of the set $(X, Y)_x$ is in $\ker \pi_* \cap H$, then $i(\xi)(d\omega) = 0$ since $d\omega = \pi^* d\beta$ and $\xi \in \ker \pi$. Moreover $i_\xi d\alpha = 0$ since $\ker \pi_* \cap H$ is an isotropic distribution (by definition of completely integrable contact structures). Hence

$$\Big((d\omega)^k \wedge (d\alpha)^{n-k}\Big)(X, Y) = 0,$$

if any element of (X, Y) is in $\ker \pi_* \cap H$.

We may now assume that each element of $(X, Y)_x$ is π-horizontal, i.e. that each vector X_x, Y_x is a combination of $\left\{\dfrac{\partial}{\partial q_i}\right\}$ $\left(\text{so does not have } \dfrac{\partial}{\partial \theta_i}\right.$ components$\Big)$. Since

$$d\alpha = \sum_{i=0}^{n} dq_i \wedge d\theta_i$$

we have that:

$$d\alpha \left(\frac{\partial}{\partial q_i}, \frac{\partial}{\partial q_j} \right) = 0$$

and therefore

$$\left((d\omega)^k \wedge (d\alpha)^{n-k} \right)(X, Y) = 0.$$

We therefore have proved that

$$\alpha_t \wedge (d\alpha_t)^n = \alpha \wedge (d\alpha)^n.$$

Hence for all $t \in [0, 1]$, $\alpha_t = \alpha + t\omega$ is a contact form.

By Gray stability theorem [18], [26] the contact structure $h = [\alpha]$ defined by the contact form $\alpha = \alpha_0$ is equivalent to the contact structure $[\alpha_1]$ defined by

$$\alpha_1 = \alpha - \pi^* \beta = \Phi^* \alpha',$$

which is equivalent to the contact structure $H' = [\alpha']$.

Hence there exist a family of functions f_t and a family of diffeomorphisms $\rho_t : P \longrightarrow P'$ such that $\rho_t^* \alpha_0 = f_t \alpha_0$. Hence

$$(\Phi \circ \rho_1)^* \alpha' = f_1 \alpha.$$

Checking the details of Martinet's proof of Gray' stability theorem [25], [26], reveals that $\pi' \circ \rho_t = \pi$. The proof of the theorem is now completed.

$\square$

B.5   Global T^{n+1} actions

B.5.1   The convexity and realization theorems

We consider now an oriented contact manifold (P, H), with a global action of T^{n+1} preserving the contact structure H.

As usual let $\pi : P \longrightarrow W$ be the canonical projection onto the orbit space W, which is a smooth manifold of dimension n with boundary and corners.

There exists a global contact form α, representing the contact structure H and which is invariant under the T^{n+1} action. Indeed if β is any contact form representing H, then for any $\tau \in T^{n+1}$, denote again by τ the diffeomorphism of P: $x \mapsto \tau \cdot x$. We obtain the invariant contact form α as:

$$\alpha = \int_{T^{n+1}} \tau^* \beta \, d(\mu(\tau))$$

where μ is the Haar measure on T^{n+1}. See [24].

In general if a Lie group G acts on a contact manifold (P, α) preserving the contact form α, one defines the moment map:

$$
\begin{aligned}
J: \quad P \quad &\longrightarrow \quad \mathcal{G}^* \\
x \quad &\longmapsto \quad J(x)(\xi) = (i(\xi_P)\alpha)(x)
\end{aligned}
$$

for all ξ in the Lie algebra $\mathcal{G}$ of G. Here ξ_P is the fundamental vector field of P defined by ξ . This moment map has the same property as the moment map of a hamiltonian action in symplectic geometry. In particular it is G-equivariant.

In the case $G = T^{n+1}$, we let $Y_0, \cdots, Y_n$ be the fundamental vector fields associated with a basis of the Lie algebra of T^{n+1}. Let $q_i = i_{Y_i}\alpha$ be the components of the moment map $J : P \longrightarrow \mathbb{R}^{n+1}$. It is clear that J factors through a map $J_W : W \longrightarrow \mathbb{R}^{n+1}$ which is an immersion, since the functions $q_i, i = 1, \cdots, n$ are the action coordinates on W. Let $K = J(P) = J_W(W) \subset \mathbb{R}^{n+1}$ denote the image of the moment map.

The main result of this chapter is the following theorem, which plays in contact geometry the role of the Atiyah-Guillemin-Sternberg convexity theorem [19] and the role of Delzant realisation theorem [13] in Symplectic Geometry.

Theorem C

Let (P, H) be a contact manifold with an effective action of T^{n+1} preserving the contact structure. Let α be a contact form representing H, which is invariant by the T^{n+1} action. Let $\pi : P \longrightarrow W$ denote the natural projection onto the orbit space W, let $J : P \longrightarrow \mathbb{R}^{n+1}$ be the moment map, factoring through $J_W : W \longrightarrow \mathbb{R}^{n+1}$ and $K \subset \mathbb{R}^{n+1}$ its image.

1. *Suppose the T^{n+1} action is regular, then:*

   (a) *W is diffeomorphic to the sphere S^n.*

   (b) *If $n \geqslant 2$, J_W is an embedding which identifies W with the hypersurface K.*

   (c) *If $n \geqslant 3$, then $P \approx T^{n+1} \times S^n$, and the image K of the moment map determines the contact structure H.*

2. *Suppose the T^{n+1} action is singular and $n \geqslant 2$, then:*

(a) *The rays in $\mathbb{R}^{n+1}$ from the origin and leaning on K generate a closed convex polytop C.*

(b) *J_W is an embedding which allows to identify W with the hypersurface with boundary and corners K.*

(c) *The image K of the moment map determines the contact structure H.*

Proof

1. The regular case:

   The results in the regular case were known to Lutz [24].

   We already have seen that J_W is transverse to lines through the origin in $\mathbb{R}^{n+1}$. Hence, composing J_W with the projection of $\mathbb{R}^{n+1} \smallsetminus \{0\}$, we obtain an etale map $W \longrightarrow S^n$. This proves the first assertion.

   For the second assertion, if $n \geqslant 2$, this etale map is a diffeomorphism.

   The third assertion follows from the classification theorem B since $H^2(W, \mathcal{R}_1) = H^2(S^n, Z^{n+1})$ is trivial for $n \geqslant 3$.

2. The singular case:

   The symplectization of a contact manifold (P, α) is the symplectic manifold $(\widetilde{P}, \widetilde{\omega})$ where $\widetilde{P} = P \times \mathbb{R}^*_+$ and $\widetilde{\omega} = -td\alpha - dt \wedge \alpha$.

   The action of T^{n+1} on the first factor of $\widetilde{P}$ is a hamiltonian action with moment map $\widetilde{J}$ such that

   $$\widetilde{J}(x, t) = tJ(x)$$

   where $J : P \longrightarrow \mathbb{R}^{n+1}$ is the contact moment map. Therefore according to the terminology of Theorem C, the image $\widetilde{K}$ of $\widetilde{J}$ is tK, $t \in \mathbb{R}^*_+$, which means that $\widetilde{K} = C \smallsetminus \{0\}$.

   The operation of multiplication on the second factor of $\widetilde{P}$ defines an action of $\mathbb{R}^*_+$ which descends to the orbit $\widetilde{W}$ of T^{n+1} and corresponds to homotheties of $\mathbb{R}^{n+1}$.

   Let us now follow the methods used by Conevaux-Dazord-Molino in [11] to prove Atiyah-Guillemin-Sternberg convexity theorem. The convexity of C will follow from the existence of geodesics in $\widetilde{W}$ for the flat metric which is the pull back of the flat euclidean metric on $\mathbb{R}^{n+1}$.

The orbit space $\widetilde{W}$ has the structure of an affine manifold with boundary and corners. It is locally convex. The mapping $\tilde{J} : \tilde{P} \longrightarrow \mathbb{R}^{n+1}$ factors through a proper embedding:

$$\tilde{J}_{\widetilde{W}} : \widetilde{W} \longrightarrow \mathbb{R}^{n+1}.$$

The manifold W is a compact hypersurface of $\widetilde{W}$ and each orbit of $\mathbb{R}^{n+1}$ in $\widetilde{W}$ meets W.

We endow $\widetilde{W}$ with the flat metric $\tilde{g}$, which is the pull-back of the standard metric on $\mathbb{R}^{n+1}$ and let $\tilde{d}$ be the corresponding distance. We denote by d the Euclidean distance on $\mathbb{R}^{n+1}$.

For any $\tilde{q}_0, \tilde{q} \in \widetilde{W}$ and $q_0 = \tilde{J}_{\widetilde{W}}(\tilde{q}_0)$; $q = \tilde{J}_{\widetilde{W}}(\tilde{q})$, we have:

$$\tilde{d}(\tilde{q}_0, \tilde{q}) \leqslant d(q_0, 0) + d(q, 0).$$

Indeed, for any $\lambda_0, \lambda \in \mathbb{R}_+^*$, we have:

$$\begin{aligned}
d(\tilde{q}_0, \tilde{q}) &\leqslant \tilde{d}(\tilde{q}_0, \lambda_0\tilde{q}_0) + \tilde{d}(\lambda_0\tilde{q}_0, \lambda\tilde{q}) + \tilde{d}(\lambda\tilde{q}, \tilde{q}) \\
&< d(q_0, 0) + d(q, 0) + \tilde{d}(\lambda_0\tilde{q}_0, \lambda\tilde{q}) \qquad \text{(B.5.1)}
\end{aligned}$$

if $\lambda_0, \lambda \leqslant 1$. Observe that the last term can be made arbitrarily small by choosing λ_0 and λ so that $\lambda_0\tilde{q}_0$ and $\lambda\tilde{q}$ belong to εW, where ε is arbitrarily small. Hence $\tilde{d}(\lambda_0\tilde{q}_0, \lambda\tilde{q}) \leqslant \varepsilon D$ where D is the diameter of W for the metric $\tilde{g}$.

Assume now the inequality (B.5.1) is strict, i.e. there exist a $\delta \geqslant 0$ such that: $\tilde{d}(\tilde{q}_0, \tilde{q}) \leqslant d(q_0, 0) + d(q, 0) - \delta$, since $\tilde{J}_{\widetilde{W}}$ decreases the lengths of paths, one can find r and R such that any path of length less or equal to $d(q_0, 0) + d(q, 0) - \dfrac{\delta}{2}$ joining $\tilde{q}_0$ to $\tilde{q}$ projects in $\mathbb{R}^{n+1}$ along a path contained in the compact shell $C_{r,R}$ which is the intersection of C and the space between the spheres of radius R and r. Let $\widetilde{W}_{r,R} \subset \widetilde{W}$ be the pre-image of $W_{r,R}$. By the usual method of successive approximations, one constructs in the compact subset $\widetilde{W}_{r,R}$ a continuous curve which is a geodesic joining $\tilde{q}_0$ to $\tilde{q}$. Since $\mathcal{W}$ is locally convex (see [x]), and since $\tilde{J}_{\widetilde{W}}$ is a local isometry, this geodesic projects in $\mathbb{R}^{n+1}$ along a segment of the same length. Hence $\tilde{d}(\tilde{q}_0, \tilde{q}) = d(q_0, q)$.

We have shown that for arbitrary points $\tilde{q}_0$ and $\tilde{q}$ projecting onto q_0 and q, we have:

either $\tilde{d}(\tilde{q}_0, \tilde{q}) = d(q_0, 0) + d(q, 0)$ or $\tilde{d}(\tilde{q}_0, \tilde{q}) = d(q_0, q)$.

Fix $\tilde{q}_0$ and let $\tilde{A} = (J_{\widetilde{W}})^{-1}\{tq_0\,;\ t \leqslant 0\}$. We partition $\tilde{A}$ into the following disjoint open subsets A_1 and A_2 where:

$$\tilde{A}_1 = \{\tilde{q} \in \tilde{A}\ \ \tilde{d}(\tilde{q}_0, \tilde{q}) = d(q_0, 0) + d(q, 0)\}, \qquad (B.5.2)$$

$$\tilde{A}_2 = \{\tilde{q} \in \tilde{A}\ \ \tilde{d}(\tilde{q}_0, \tilde{q}) = d(q_0, q)\}. \qquad (B.5.3)$$

In $n \geqslant 2$, then $\tilde{A}$ is connected, and since $\tilde{A}_2$ is a neighborhood of $\tilde{q}_0$, $\tilde{A} = \tilde{A}_2$. Therefore $\widetilde{W}$ is convex and $\tilde{J}_{\widetilde{W}}$ is a global isometry. This proves the first statement of Theorem C.

In particular, $J_W : W \longrightarrow \mathbb{R}^{n+1}$, the restriction of $\tilde{J}_{\widetilde{W}}$ to W is an embedding which identifies W with the hypersurface with boundary and corners K. The trace K_0 of C on the unit sphere in $\mathbb{R}^{n+1}$ is contractible and the radial projections etales $(W, \partial W)$ onto $(K_0, \partial K_0)$. This etale map is a diffeomorphism and hence each half line from the origin in C meets K in a single points. This proves the assertion (b).

The last assertion results from the classification Theorem B and the fact that K is contractible. $\qquad \square$

Remark B.4

1. *If we view $\mathbb{R}^{n+1}$ as the Lie algebra of T^{n+1}, the hyperplanes which form the boundary of C correspond to the equations of the Lie algebras of the 1-dimensional isotropies of the torus action. Therefore, they are rational hyperplanes and through each vertex of C pass n hyperplanes. We will call such a cone a "rational polyhedral cone".*

2. *It is easy to check that if K is a compact hypersurface with boundary and corners in $\mathbb{R}^{n+1}$ with the following properties:*

   (a) *K does not contain the origin and is transverse at each point q to the line through q and the origin.*

   (b) *The half lines through the origin meeting K form a rational polyhedral convex cone, then K is the image of the contact moment map of a T^{n+1} action preserving the contact form of a compact contact manifold of dimension $2n + 1$.*

B.5.2 Miscellaneous and applications to K-contact structures

Consider now a contact manifold (P, α), where the contact form α is invariant under an effective action of T^{n+1}. We have seen that the Reeb field Z of α is a linear combination of the fundamental vector fields $Y_0, Y_1, \cdots, Y_n$ of the T^{n+1} action:

$$Z = \sum_{i=0}^{n} Z_i Y_i$$

where Z_i are basic functions. Recall that $q_i = i(\alpha)Y_i$ are the components of the moment map and $\{j^1 q_0, \cdots, j^1 q_n\}$ form the local basis of the Legendre lattice $\mathcal{R}_1$.

We say that Z is a fundamental vector field of the action if Z is a combination of Y_i with constant coefficients $\lambda_= Z_i$. In that case the image K of the moment map will be contained in the affine hyperplane H satisfying the equation:

$$\sum_{i=0}^{n} \lambda_i q_i = 1$$

we call the **Reeb hyperplane**.

Observe that in this case, the action cannot be regular, since there is no etale mapping from the n-sphere S^n onto a affine space of the same dimension.

We will have: $K = H \cap C$ and hence that K itself is a degenerate convex polytope. Therefore we get the following:

Proposition B.10

Assume the Reeb field is a fundamental vector field of the T^{n+1} action, then:

1. *The image K of the moment map is the intersection of an affine hyperplane H which does not contain the origin with a rational polyhedral convex cone C in $\mathbb{R}^{n+1}$.*

2. *Conversely, such a hyperplane and such a cone determine a $(2n + 1)$-dimensional compact contact manifold (P, α), an effective action of T^{n+1} on P preserving the contact form, and such that its Reeb field is a fundamental vector field of this action. Moreover the contact structure H determined by α depends only on $K = H \cap C$.*

Application to compact K-contact manifolds

Recall that a contact manifold (P, H), with $H = [\alpha]$, where α is contact form with Reeb field Z is said to be a K-contact manifold if Z is a Killing vector field for some contact metric g. Hence the flow of Z is a 1-parameter group of isometries of P. Suppose now that P is compact.

The classical theorem of Meyer-Steenrod, asserts that the group $I(M)$ of isometries of a compact riemannian manifold is a compact Lie group.

Consider now our compact K-contact manifold $(P, [\alpha])$. Then the flow ϕ_t of the Reeb field is subgroup of the compact Lie group $I(P)$. Its closure G in $I(P)$ is a compact abelian group, hence a torus T_k. Therefore, if $s \in G$, we have that $s \circ \phi_t = \phi_t \circ s$ and thus $s_* Z = Z$ since G is abelian. Consequently:

$$
\begin{aligned}
s^* \alpha(x)(X) &= \alpha(s(x))(s_* X) \\
&= g(s(x))\big(Z(s(x)), s_* X(x)\big) \\
&= (s^* g)(Z, X)(x) = g(x)(Z, X) = \alpha(X). \quad \text{(B.5.4)}
\end{aligned}
$$

The torus action thus preserves the contact form α. The dimension of G is a number between 1 and $n+1$ if P is $2n+1$ dimensional.

The case when $G = T^1$ is well known: those manifolds have been studied by Thomas [41], who called them "almost regular" contact manifolds, generalizing "regular contact manifolds" studied by Boothby and Wang [9].

Here we consider the opposite case in which $G = T^{n+1}$, and say that the K-contact manifold (P, α, g) is of **non-degenerate completely integrable type**. We are in the situation described above: the Reeb field Z is a linear combination of the fundamental vector fields of the action with constant coefficients; but this time these coefficients must be rationally independent. Hence we may apply the proposition above to non-degenerate completely integrable K-contact manifolds and state an existence and uniqueness theorem like in the above proposition.

References

1. Arnold V., *Mathematical Methods of Classical Mechanics*, Springer Graduate Texts in Mathematics, **60**, (1978).

2. Arnold V., Gusein, Zade and Varchenko, *Singularities of Differentiable maps*, Birkhauser, (1985).

3. Atiyah M., *Convexity and commuting hamiltonians*, Bull. Lond. Math. Soc., **14**, (1982), pp. 1–15.

4. Audin M., *Hamiltoniens périodiques sur les variétés symplectiques compactes de dimension* 4, Springer Lecture Notes in Mathematics **1416**, (1990), pp. 1–25.

5. Banyaga A., *On characteristics of hypersurfaces in symplectic manifolds*, Proceed. Symposia in Pure Mathematics **54**, Part 2, (1993), pp. 9–17.

6. Banyaga A. and Molino P., *Géometrie des formes de contact complétement intégrable de type torique*, Séminaire Gaston Darboux, Montpellier, **92** (1991), pp. 1–25.

7. Banyaga A. and Rukimbira P., *An invitation to Contact Geometry*, Preprint.

8. Blair D. E., *Contact manifolds in riemanian geometry*, Springer lecture Notes in Math., **509**, (1979).

9. Boothby A. M. and Wang H. C., *On contact manifolds*, Ann. of Math., **60** (1978), pp. 721–734.

10. Boucctta M. and Molino P., *Géometrie globale des systèmes complétement integrables: fibrations lagrangiennes singulières et coordonnées action-angle à singularités*, C. R. Acad. Sc. Paris, **308**, 1, (1989), pp. 421–424.

11. Condevaux M., Dazord P. and Molino P., *Géometrie du moment*, Séminaire Rhodanien, Lyon, (1978).

12. Dazord P. and Delzant T., *Le problème général des variables action-angle*, Jour. of Differ. Geom., **26**, 2, (1987), pp. 223–252.

13. Delzant T., *Hamiltoniens périodiques et image convexe de l'application moment*, Bull. Soc. Math. France, **116**, (1988), pp. 315–339.

14. Desolneux-Moulis N., *Singular lagrangian foliations associated to an integrable hamiltonian vector field*, In Symplectic Geometry, groupoids and integrable systems. MSRI Publication no 20, Springer Verlag (1991), pp. 129–136.

15. Dufour J. P. and Molino P., *Compactification des actions de $\mathbb{R}^n$ et variables action-angle avec singularités*, In Symplectic Geometry, groupoids and integrable systems. MSRI Publication no 20, Springer Verlag, (1991), pp. 151–168.

16. Duistermaat J. J., *On global action-angle variables*, Comm. on Pure and Appl. Math., **33**, (1980), pp. 687–706.

17. Eliasson H., *Normal forms for hamiltonian systems with Poisson commuting integrals*, Comment. Math. Helv., **65**, 1, (1990), pp. 4–35.

18. Gray J. W., *Some global properties of contact structures*, Ann. of Math., **69**, (1959), pp. 421–450.

19. Guillemin W. and Sternberg S., *Convexity properties of the moment mapping I*, Invent. Math., **67**, (1982), pp. 491–513.

20. Kostant B., *Quantization and representation theory I*, Springer Lecture Notes in Math., **170**, (1970), pp. 87–208.

21. Koszul J. L., *Sur certains groupes de transformation de Lie*, Colloque de Géometrie Différentielle, Strasbourg, (1953).

22. Liberman P., *Legendre foliations on contact manifolds*, Differential Geometry and its Applications, **1**, (1991), pp. 57–76.

23. Liberman P. and Marle C. M., *Symplectic Geometry and Analytical Mechanics*, D.Reidel Publishing Co. (1987).

24. Lutz R., *Sur la géometrie des structures de contact invariantes*, Annales de l'Institut Fourier, **29**, 1, (1979), pp. 283–300.

25. Martinet J., *Sur les singularités des formes différentielles*, Annales de l'Institut Fourier, **20**, (1970), pp. 95–178.

26. Martinet J., *Formes de contact sur les variétés de dimension trois*, Springer Lecture Notes in Math, **209**, (1971), pp. 142–163.

27. Molino P., *Dualité symplectique, feuilletages et géometrie du moment*, Publications Math., **33**, (1989), pp. 533–541.

28. Molino P., *Du theórème d'Arnold-Liouville aux formes normales de systèmes hamiltoniens toriques: une conjecture*, Séminaire Gaston-Darboux, Montpellier, (1989-90).

29. Monna G., *Feuilletages de K-contact sur les variétés compactes de dimension trois*, Publications Math. UEB, **30**, (1984), pp. 81–87.

30. Moser J., *On the volume element of a manifold*, Trans. AMS, **120**, (1965), pp. 286–294.

31. Reeb G., *Sur certaines propriétés topologiques des trajectoires des systèmes dynamiques*, Mem. Ac. Roy. Belg., **27**, (1952), pp. 130–194.

32. Pang M. Y., *The structure of Legendre foliations*, Thesis, Univ. of Washington (1989).

33. Reinhart B., *Foliated manifolds with bundle-like metrics*, Ann. of Math., **69**, (1959), pp. 119–132.

34. Satake I., *The Gauss-Bonnet theorem for V-manifolds*, Jour. Math. Soc. Japan, **9**, (1957), pp. 464–492.

35. Sasaki S., *Almost contact manifolds III*, Lecture Notes Tohoku Univ., Japan (1968).

36. Schwartz G., *Lifting smooth homotopies of orbit spaces*, Publications IHES, **51**, (1980), pp. 37–136.

37. Souriau J. M., *Structure des systemès dynamiques*, Dunod, Paris (1969).

38. Spencer D., *Deformations of structures on manifolds defined by transitive continuous pseudogroups I*, Ann. of Math., **76**, (1962), pp. 306–445.

39. Spencer D., *Deformations of structures on manifolds defined by transitive continuous pseudogroups II*, Ann. of Math., **81**, (1965), pp. 389–450.

40. Sussmann H., *Orbits of families of vector fields and integrability of distributions*, Trans. AMS, **180**, (1973), pp. 171–178.

41. Thomas C. B., *Almost regular contact manifolds*, J. Diff. Geom., **11**, (1978), pp. 521–533.

42. Weinstein A., *Periodic orbits for convex hamiltonian systems*, Ann. of Math., **108**, (1978), pp. 507–518.

Bibliography

[Abr-Mar67] R. Abraham and J. E. Marsden, *Foundations of mechanics*, Benjamin, NY, (1967).

[Arn78] V. I. Arnold, *Mathematical methods of classical mechanics*, Springer Graduate texts in Mathematics, Springer, **60** (1978).

[Arn65] V. I. Arnold, *Sur les propriétés topologiques des applications globalement canoniques de la mécanique classique*, C.R Acad. Sci. Paris (1965), pp. 3719-3722.

[Arn89] V. I. Arnold, *Conatact geometry and wave propagations*, Monograph no 34, Enseig. Math. (1989), Univ. of Geneva.

[Ati82] M. F. Atiyah, *Convexity and commuting Hamiltonians*, Bull. London Math. Soc. **14** (1982), no 1, pp. 1-15.

[Ban75] A. Banyaga, *Sur le groupe des difféomorpismes qui préservent une formes de contact régulière*, C. R. Acad. Sc. Paris **281** (1975), serie A, pp. 707-709.

[Ban78] A. Banyaga, *Sur la structure des difféomorphismes qui préservent une forme symplectique*, Comment. Math. Helv. **53** (1978) pp. 174-2227.

[Ban80] A. Banyaga, *On fixed points of symplectic maps*, Inventiones Math. **5** (1980), pp. 215-229.

[Ban86] A. Banyaga, *On isomorphic classical diffeomorphism groups I*, Proceed. Amer. Math. Soc. **98** (1986), no 1, pp. 113-118.

[Ban88] A. Banyaga, *On isomorphic classical diffeomorphism groups II* J. Differ. Geom. **28** (1988), pp. 93-114.

[Ban97] A. Banyaga, *The structure of classical diffeomorphism groups*, Mathematics and its applications no 400, Kluwer Academic Publisher, (1997).

[Ban99] A. Banyaga, *The geometry surrounding the Arnold-Liouville theorem*, Advances in Geometry, Progress in Math. **172**, Birkhauser Boston (1999), pp. 53-69.

[Ban90] A. Banyaga, *A note on Weinstein's conjecture*, Procced. AMS **109** (1990), no 3, pp. 855-858.

[Ban10] A. Banyaga, *A Hofer-like metric on the group of symplectic diffeomorphisms*, Symplectic topology and mesure preserving dynamical systems 1-23, Contemp. Math. **512** (2010), Amer. Math. Soc. Providence R.I.

[Ban-Don06] A. Banyaga and P. Donato, *Length of contact isotopies and extensions of the Hofer metric*, Ann. Global Analy. Geometry **30** (2006), no 3, pp. 299-312.

[Ban-Ine95] A. Banyaga and A. Mc Inerney A., *On isomorphic classical diffeomorphism groups III*, Annals of Global Analysis and Geometry **13** (1995), pp. 117-127.

[Ban-Tch14] A. Banyaga and S. Tchiuaga *The group of strong homeomorphisms in the L^∞-metric*, Adv. Geom. **14** (2014), no 3, pp. 523-529.

[Ban-Spa] A. Banyaga and P. Spaeth, *Uniqueness of contact hamiltonians of topological strict contact isotopies*, preprint.

[Ban-Hur-Spa16] A. Bayaga, D. Hurtubise and P. Spaeth, *The symplectic displacement energy*, J. Symp. Geo. (2016)

[Ben83] D. Bennequin, *Entrelacement et equation de Pfaff*, Astérisque **107-108** (1983), pp. 87-162.

[Bla02] D. E. Blair, *Riemannian geometry of contact and symplectic manifolds*, Birkhäuser Boston, Inc., Boston (2002).

[Boo69] W. M. Boothby, *Transitivity of automorphisms of certain geometric structures*, Trans. Amer. Math. Soc. **137** (1969), pp. 93-100.

[Boo-Wan78] W. M. Boothby and H. C. Wang, *On contact manifolds*, Ann. Math. **60** (1978), pp. 721-734.

[Bot-Tu82] R. Bott and L. V. Tu, *Differential forms in Algebraic topology*, Graduate Text in Math. **82** (1982), Springer New-York, Berlin.

[Bou74] F. Bourgeois, *Odd dimensional tori are contact manifold*, Int. Math. Res. **2002** (1574), no 30.

[Bou02] F. Bourgeois, *A Morse-Bott approach to contact homology*, thesis, Stanford (2002).

[Buh14] L. Buhovsky, *Variation on Eliashberg-Gromov theorem I*, C^0 symplectic Topology and Dynamical Systems (2014), IBS Center for Geometry and Physics, Korea, Lecture.

[Buh-Sey13] L. Buhovsky and S. Seyfaddini, *Uniqueness of generating Hamiltonians for topology Hamiltonian flows*, J. Symplectic Geom. **11** (2013) no 1, pp. 37-52.

[Cal70] E. Calabi, *On the group of automorphisms of a symplectic manifold*, Problems in analysis, a symposium in honor of S. Bochner, pp. 1-26, Princeton University Press, Princeton (1970).

[Car-Vit08] F. Cardin, C. Viterbo, *Commuting Hamiltonian and multi-time Hamilton-Jacobi equations*, Duke Math. J. **144** (2008), pp. 235-284.

[Dui80] J. J. Duistermaat, *On global action-angle coordinates*, Comm. on Pure and Applied Math. **33** (1980), pp. 687-706.

[Del88] J. Delzant, *Hamiltoniens périodiques et image convexe de l'application moment*, Bull. Soc. Math. France **116** (1988), pp. 129-136.

[Eli87] Y. Eliashberg, *A theorem on the structure of wave fronts and its application in symplectic topology* Funct. Anal. and Its Applications **21** (1987), pp. 227-232.

[Eli-Pol93] Y. Eliashberg and L. Polterovich, *Bi-invariant metrics on the group of Hamiltonian diffeomorphisms*, Inter. J. Math. **4** (1993), no 5, pp. 727-730

[Ent-Pol10] M. Entov, L. Polterovich, C^0-*rigidity of the Poisson bracket*, Contemporary Math. **512** (2010), pp. 25-32.

[Ent-Pol-Zap07] M. Entov, L. Polterovich, F. Zapolsky, *Quasi-morphisms and Poisson bracket*, Pure and Applied Math. Quarterly **3-4** (2007), pp. 1037-1055.

[Gir94]   E. Giroux, *Une structure de contact, même tendue est plus ou moins tordue*, Ann. Scient. Ec. Norm. Sup. **27** (1994), pp. 697-705.

[Gra59]   J. W. Gray, *Some global properties of contact structures*, Ann. Math. **64** (1959), pp. 421-450.

[Gro86]   M. Gromov, *Soft and hard symplectic geometry*, Proceedings of the International Congress of Mathematicians 1 (Berkeley, CA, 1986 pp. 81-98), Amer. Math. Soc. Providence, (1987).

[Gro85]   M. Gromov, *Pseudoholomorphic curves in symplectic manifolds*, Invent. Math. **82** (1985), pp. 307-347.

[Gro86]   M. Gromov, *Partial Differential Relations*, Ergebnisse der Mathematik, Springer (1986).

[Gui-Ste77]   V. W. Guillemin and S. Sternberg, *Geometrics asymptotic*, AMS, Providence, RI, (1977).

[Gui-Ste90]   V. W. Guillemin and S. Sternberg, *Symplectic techniques in physics*, Cambridge University Press, Cambridge (1990).

[Hir76]   M. Hirsch, *Differential Topology*, Graduate Text in Mathematics, no 33 Springer-Verlag, New York-Heidelberg **3** (1976), corrected reprint (1994).

[Hof90]   H. Hofer, *On the topological properties of symplectic maps*, Proceed Royal Soc. Edinburg **115** (1990), pp. 25-38

[Hof-Zeh94]   H. Hofer, E. Zehnder, *Symplectic invariant and Hamiltonian dynamics*, Birkhaüser, (1994).

[Hum08]   V. Humilière, *Continuité en topologie symplectique*, thesis, Ecole Polytechnique (2008).

[Hum09]   V. Humilière, Hamiltonian pseudo-representation, Comm. Math. Helv. **84** (2009), pp. 571-585.

[Kir76]   A. Kirilov, *Éléments de la théorie des représentations*, Edition Mir, Moscou, (1976).

[Kob72] S. Kobayashi, *Transformation groups in differential geometry*, Erg. Math. Grenzbeg **70**, Springer-Verlag, (1972).

[Lal-McD95] F. Lalonde and D. McDuff, *The geometry of symplectic energy*, Ann. of Math. **2**, (1995), no 349-371.

[Lib-Mar87] P. Liberman and C-M. Marle, *Symplectic geometry and Analytical Mechanics*, D. Reidel Publishing Co., Dordrecht (1987).

[Mar70] J. Martinet, *Sur les singularités des formes différentielles*, Ann. Institut Fourier, Grenoble **20** (1970) pp. 95-178.

[Mar70] J. Martinet, *Formes de contact sur les variétés de dimension 3*, Proceed. of Liverpool Singularities Symposium II (1969/1970), pp. 142-163, Lectures notes in Math. **209**, Springer, Berlin (1971).

[McD-Sal95] D. McDuff and D. Salamon, *Introduction to symplectic topology*, Oxford Mathematical Monographs, Oxford University Press, New York (1995).

[Mos65] J. Moser, *On the volume elements of a manifold*, Trans. Amer. Math. Soc. **120** (1965) pp. 286-294.

[Mül-Spa15] S. Müler and P. Spaeth *Topological contact dynamics I: symplectization and applications of the energy-capacity inequality*, Adv. Geom. **15** (2015) no 3, pp. 349-380.

[Mül-Spa14] S. Muler and P. Spaeth *Topological contact dynamics II: topological automorphisms, contact homeomorphisms and non-smooth dynamical systems*, Trans. Amer. Math. Soc. **366** (2014) no 9, pp. 5009-5041.

[Mül-Spa] S. Muler and P. Spaeth *Topological contact dynamics III, Uniqueness of the topoligical hamiltonian*, preprint.

[Oh-Mül07] Y-O Oh and S. Muller, *The group of hamiltonian homeomorphisms and the C^0- symplectic topology*, J. Symplectic Geom. **5** (2007), pp. 167- 225.

[Pol01] L. Polterovich, *On the geometry of the group of symplectic diffeomorphisms*, lectures in Mathematics, ETH Zurich, Birkhauser Basel (2001).

[Pol-Dan14] L. Polterovich and R. Daniel, *Functions theory on symplectic manifolds* CRM Monograph Series, bf 34, AMS Providence R.I (2014).

[Ruk95] P. Rukimbira, *Topology and closed characteristics of K-contact manifolds*, Bull. Belg. Math. Soc. (1995) pp. 349-356.

[Ryb10] T. Rybicki, *Commutators of contactomorphisms*, Adv. Math **225** (2010), no 6, pp. 3291-3326.

[Sil01] A. C. da Silva, *Lectures on symplectic geometry*, Springer Lecture Notes in Mathematics **1764** (2001).

[Ste64] S. Sternberg, *Lectures on differential geometry*, Prentice Hall, Englewood Cliffs (1964).

[Tak-Dum73] F. Takens, F. Dumortier, *Characterization of compactness of symplectic manifolds*, Bol. Soc., Brasil. Mat. **4** (1973), pp. 167-173.

[Thu76] W. P. Thurston, *Some simple examples of symplectic manifolds*, Proc. A.M.S **55** (1976), pp. 467-468.

[Vit87] C. Viterbo, *A proof of Weinstein conjecture on* $\mathbb{R}^{2n}$, Ann. Inst. Poincaré, Anal. Non Linéaire **4** (1987), pp. 337-356.

[Vit06] C. Viterbo, *On uniqueness of generating Hamiltonians for continuous limits of Hamiltonian flows*, Intern. math. Res. Notes (2006), ArtID34028 **9**, Erratum ID 38784, **4**.

[War71] F. Warner, *Foundation of differentiable manifolds and Lie groups*, Scotl., Foresman and Co., London (1971).

[Wei77] A. Weinstein, *Lectures on symplectic manifolds*, CBMS Regional Conf. Series in Math. **29** (1977), Amer. Math. Soc., Providence.

[Wei71] A. Weinstein, *A symplectic manifolds and their Lagrangian submanifolds*, Advances in Math. **6** (1971) pp. 329-245.

[Wei81] A. Weinstein, *Symplectic Geometry*, Bull. Amer. Math. Soc. **5** (1981) no 1, pp. 1-13.

[Wei79] A. Weinstein, *On the hypotheses of Rabinowiz periodic orbit theorems*, J. Differential Equations **33** (1979) no 3, pp. 353-358

Index

Printed in the United States
By Bookmasters